Legal Disclaimer

Copyright 2017- All rights reserved.

In no way is it legal to reproduce, duplicate, or transmit any part of this document in either electronic means or in printed format. Recording of this publication is strictly prohibited and any storage of this document is not allowed unless with written consent by the publisher. All rights reserved.

The information provided herein is stated to be truthful and consistent, in that any liability, in terms of inattention or otherwise, by any usage or abuse of any policies, processes, or directions contained within is the sole and utter responsibility of the recipient. Under no circumstances will any legal responsibility or blame be held against the publisher for any reparation, damages, or monetary loss due to the information herein, either directly or indirectly.

Respective author(s) own all copyrights not held by the publisher.

Legal Notice:
This book is copyright protected. This is only intended for personal use. You may not amend, distribute, sell, use, quote or paraphrase any part or the content within the book without consent of the author or copyright owner. Legal action will be pursued if the previous are ever breached.

Disclaimer:
Please note the information contained within this book is for educational and entertainment purposes only. Every attempt has been made to provide accurate, up to date and reliable complete information. Under no circumstances does this book express or imply any warranties of any kind. Readers must acknowledge that the author is not engaged in the rendering of any form of legal, financial, medical or professional advice.

By reading this document, the reader agrees that under no circumstances can we be held responsible for any losses, direct or indirect, which are incurred as a result of the use of information contained within this document, including, but not limited to- errors, omissions, or inaccuracies.

Everything Electrical: How To Test Circuits Like A Pro: Part 2

Preface:

Have you ever studied electricity in a college class or trade school and still felt puzzled at the end leaving you feeling like the teacher failed you or that the theory just didn't give you anything useful to use on the field? That they didn't prepare you for those uncommon or intermittent electrical issues that leave you feeling like you don't have a plan of attack. Well either way great, you're not alone. I myself read about 10 full textbooks on electrical, electronics, industrial electrical and automotive electricity, that by the way were not cheap averaging in cost around 150$ each. But these books still left me feeling like they failed in many aspects of teaching real world electrical tips and tricks.

I write this book to educate in a simpler way for everyone to understand, beginners and veteran technicians alike. There is no reason to complicate things with big words that usually are left unexplained by other books and make it even harder to understand with bad examples. This book is priced low but because I feel that everyone should know at least the basics. I will also include many examples of each topic for better understanding. I recommend you reading the book front to back even if you feel you've read too much theory of electricity already. My goal is to make you "the electrical guy" that will fearlessly tackle any job. If this book series "Everything Electrical" does not teach you everything you wanted to know, I guarantee that it will at least be a very powerful supplement to your learning on electrical testing at a low price. Thank you and I hope you enjoy.

My Own Take on Electricity

Everyone should know that anything technical, including electricity, involves a lot of complicated physics, and if I wanted to, I can talk about how things really happen down to the atomic level. But for the sake of keeping it as simple as I can so I can teach only what you need to know and get to working on electrical problems as soon as possible, just accept that the way I'm explaining it is just for your ease. Otherwise this book would be way too long and you would NOT want to read it all. But my methods and examples do teach and WILL work in the real world for real life electrical issues.

This book is part2 of the series on how to use your meter like a professional electrician and/or technician.

Table Of Contents:

Ch.1: Important Things To Remember Before Starting..*(1-4)*

Ch.2: Introduction To Intermittent/Random Electrical Problems...*(5-12)*

Ch.3: Vibration-Related Intermittent Issues............*(13-31)*

Ch.4: Temperature-Related Intermittent Issues..*(32-39)*

Ch.5: Relay Involved Intermittents..........................*(40-51)*

Ch:6: Miscellaneous Tips And Tricks.......................*(52-55)*

Ch. 1: Important Things To Remember Before Starting

Introduction To Part II:

In the previous "How To Test Circuits Like A Pro", we looked into how to use the voltmeter to find two common electrical problems. **Open Circuits** and **Unwanted Resistance** problems. There we saw the troubleshooting techniques that will help to find the problem and also help identify the kind of problem you are dealing with. You must now remember those skills in order to continue on with this book.

In the Part 2 of this series we will continue to add on to your electrical knowledge and teach more on how you can solve the more complex circuit problems that can exist. Let us begin right away with a quick review on the things you need to know..

Ideal V.S. Real-Life Measurements:

In the last book we gave examples with measurements that we would **ideally** prefer to see throughout a circuit. Remember that in real life we won't always see these perfect measurement numbers. Even the power source won't output the exact voltage labeled on it. Here is a quick refresher on ideal measurements vs. real life measurements.

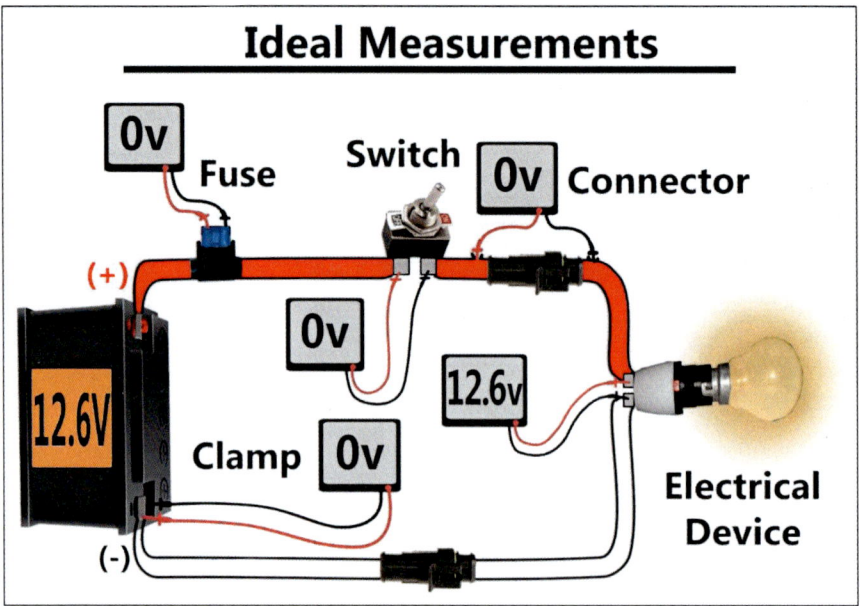

(This image shows all the **IDEAL** voltages that we would want to see across various parts in the circuit. I have added a voltmeter to each part being tested. This image also shows how, ideally, we would want ALL the voltage from the power source to be available at the electrical device, this case the light bulb. In a perfect world all the voltage from the power source would be present at the electrical load without any voltage being lost across the rest of the parts in the circuit.)

Ideally we want to have voltage measurements of 0 volts across any Switch, Fuse, Wire, Connector, Terminal, Contact or Any Other Electrical Connections. The fact is, depending on various factors, the voltage loss across these parts will increase.

The truth is that these voltage measurements shown in the previous illustration are only "ideal" measurements and are almost never the case when it comes to real life measurements. In real life, we WILL always have a small voltage drop across every electrical connection. The following is an illustration of the same circuit only now with real life measurements that we may actually encounter when testing.

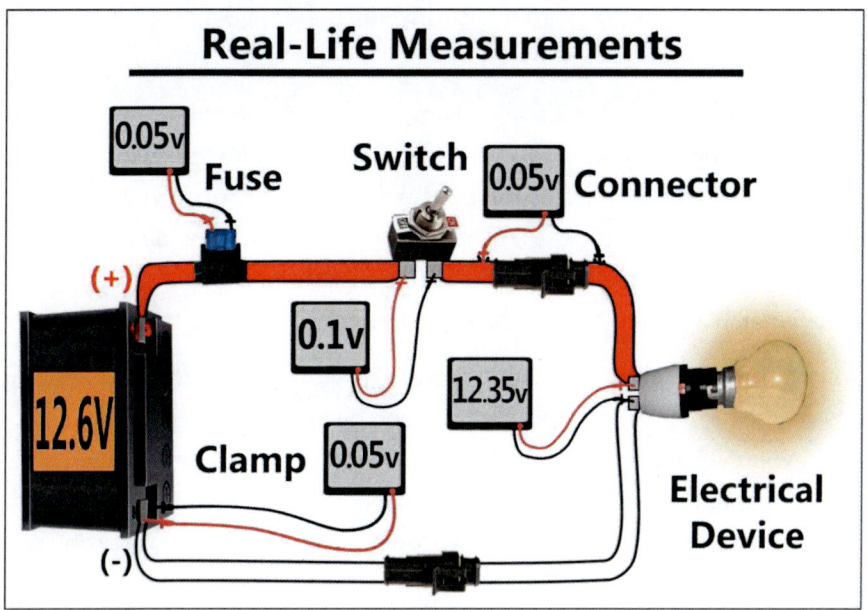

(This image shows the same basic electrical circuit as before, but only now with a **REAL** set of measurements. Notice how we do have a voltage drop, even if it is a tiny amount, across every electrical device when the circuit is ON. All the voltage from the power source will not be at the electrical device with the circuit ON)

Reminder: No voltage drop can happen without electricity flowing in the circuit. To test for voltage drops, always remember to have the circuit switch turned ON.

We may have less voltage drop or more voltage drop depending on various things that can affect a circuit. Lets look at the factors that can affect a real life measurement value..

Temperature and Other Causes of Test Measurement Variations:
When the temperature in a circuit goes up, the resistance of EVERYTHING in the circuit goes up. When the temperature in a circuit goes down, the resistance of EVERYTHING in the circuit goes down. The way this resistance change happens involves complex physics you don't really want me to explain. For the sake of simplicity, just always remember when performing electrical testing that resistance of everything in a circuit will go up if the circuit is exposed to heat and the resistance of everything in the circuit will go down if the circuit is exposed to cold temperatures.

Well why is this even important? Well when gathering your measurements, we must take this factor into account. **When a large amount of electricity is flowing through a circuit, it creates heat. Heat creates Resistance and Resistance creates voltage losses throughout the circuit**. This is why the lesson matters.

In circuits with a large amount of amperage flowing, the heat from the electricity will create voltage drops across the circuit that can actually be considered as normal. If we have a circuit with very low amperage or electricity flowing, then it will not create much heat and the voltage drops would be so tiny that they wouldn't even be noticeable. But whenever you have a circuit with a larger amperage or large amounts of electricity flowing, then you WILL have to pay attention to this rule when measuring voltage to avoid possibly confusing measurements.

When we have a large amperage, let's say 30 amps and up for example, we will have bigger voltage drops created by the heat of electricity flowing. This voltage drop will be across ALL the connections, switches, terminals, contacts, fuses, etc. This drop in voltage is a normal side effect and should not be thought of as a problem with the circuit. Instead of a perfect 0 volt reading across an electric switch, we might notice a 0.2 volt or 0.3 volt drop. Higher amperage circuit will have even higher voltage drops across parts of the circuit.

As a result of the drops in voltage we won't see all the voltage we should have at the electrical device because of this normal drop caused by high amperage flow. For example, in a 28 volt circuit you may get a reading of 27.4v or 27v at the electrical device. This isn't so much a problem, depending on how high the amperage that is flowing in the circuit. The trick is to be able to determine what **IS** a potential problem and what **IS NOT** a problem.

Important Note: Other variations to the numbers in this guide will be caused by things like..

- The amount of amperage in the circuit.
- The kind of conductor material the electric parts and wiring are made of.
- The length and type of the electrical wire used in the circuit.
- The accuracy of your testing meter.
- Temperature of the circuit.

General Ballpark Guide To Good Voltage Measurements:

When voltage testing we need to have an idea of what a real life good measurements is going to look like. The following guide will serve as only a ballpark guide to what is considered a "good" measurement. This is NOT an end all guide and there are many more specific guides out there that should be followed depending on your electrical application. But for the sake of having a general range to note what is considered "good" I have put this guide in place for that very reason.

The real trick is to understand that these "normal" voltage drops should be spread out evenly throughout the circuit. The more voltage drop in only one part of the circuit, the more problematic it may be. Also, in a circuit with medium to higher amperage, larger voltage drops will be acceptable.

The following is a general guide we can use when troubleshooting…

Low Power Low Amperage Circuits (1-10 amps):

Near 0 volts drop. Acceptable Up to a 0.5 volt drop total evenly spread out in the entire circuit.

Medium Power Medium Amperage Circuits (10-30 amps):

Minus **5%** of the Power Source's voltage.

Example 1: If power source is 12.6 volts, - 5% equals a 0.6 volt drop maximum in the entire circuit. You should not have more than 0.6 volt drop across any one place. This voltage drop should be spread out throughout the circuit..

Example 2: If power source is 120 volts, - 5% equals a 6 volt drop maximum in the entire circuit. You should not have more than 6 volt drop across any one place. Again, this voltage drop should be spread out through the circuit.

High Power High Amperage Circuits (30-100amps):

Minus **8%** of the Power Source voltage.

Example 1: If power source is 12.6 volts, - 8% equals a 1.1 volt drop maximum in the entire circuit.

Example 2: If power source is 120 volts, - 8% equals a 9.6 volt drop maximum in the entire circuit.

Very High Power High Amperage Circuits: (100-250amps):

Minus **12%** of the Power Source voltage.

Example 1: If power source is 12.6 volts, - 12% equals a 1.6 volt drop maximum in the entire circuit.

Example 2: If power source is 120 volts, - 12% equals a 14.4 volt drop maximum in the entire circuit.

Recap: Remember that you do not want to have all the voltage dropping across any one place. The voltage drop should be spread out fairly evenly throughout the circuit if it is to be considered normal voltage drops. No switch or connections should have a very large amount of the total voltage drop in the circuit across it.

Now that we have reviewed these important real world testing facts, let us begin the book...

Ch.2: Introduction To Intermittent/Random Electrical Problems:

In this chapter, we will pick up where we left off from our troubleshooting in the previous book. The third most common type of electrical problem is an **intermittent** electrical problem. These are those nasty electrical issues that only happen randomly and are not always causing a problem to the circuit. These problems may only occur for a split second to a few minutes and then disappear completely. These kind of issues are often hard to diagnose and many times hard to even confirm exists because the problem could be anywhere and it doesn't happen all the time.

The truth about these intermittent problems is that they can be seen as a kind of Open Circuit. The difference is that they cut OFF randomly and then begin working again unpredictably. Let's see a visual representation of an intermittent problem..

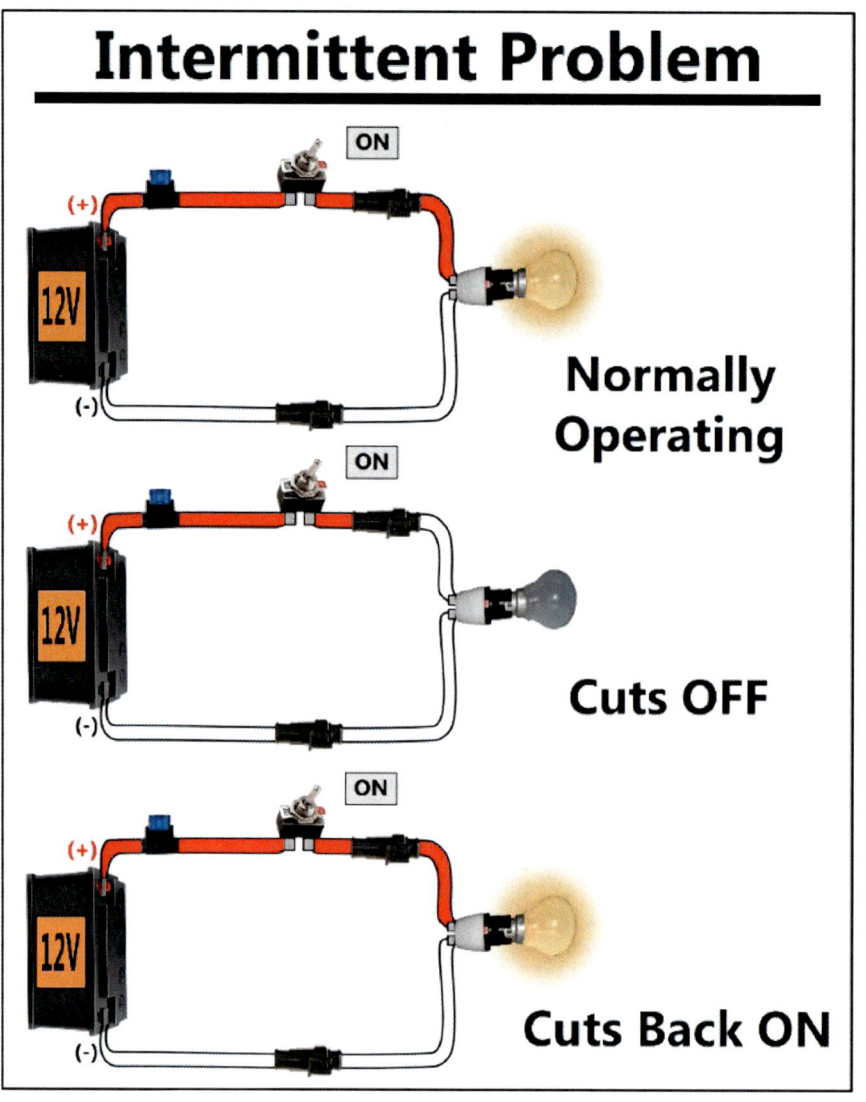

(Example of a basic intermittent electrical problem. The circuit works normally, then suddenly cuts OFF and after a few minutes and then cuts back ON again. The circuit seems to be turning ON and OFF randomly without a probable cause.)

Why it happens:

Intermittent issues are always caused when a part of the circuit loses electrical contact just enough to be able to make or break the circuit easily. These problems often happen at areas where there may be excess vibrations or excess heat around the circuit.

There are many areas as to where these problems can happen but the most common areas are usually at the contacts and terminals of the circuit. In the following we will see examples of how some intermittent issues are created...

Intermittent Connector Terminals:

One of the most common possible causes of intermittent electrical issues is a **connector**. In order to understand why these are potential intermittent problem areas, we have to understand how a connector is built. Let's take a quick look at how a connector looks disassembled...

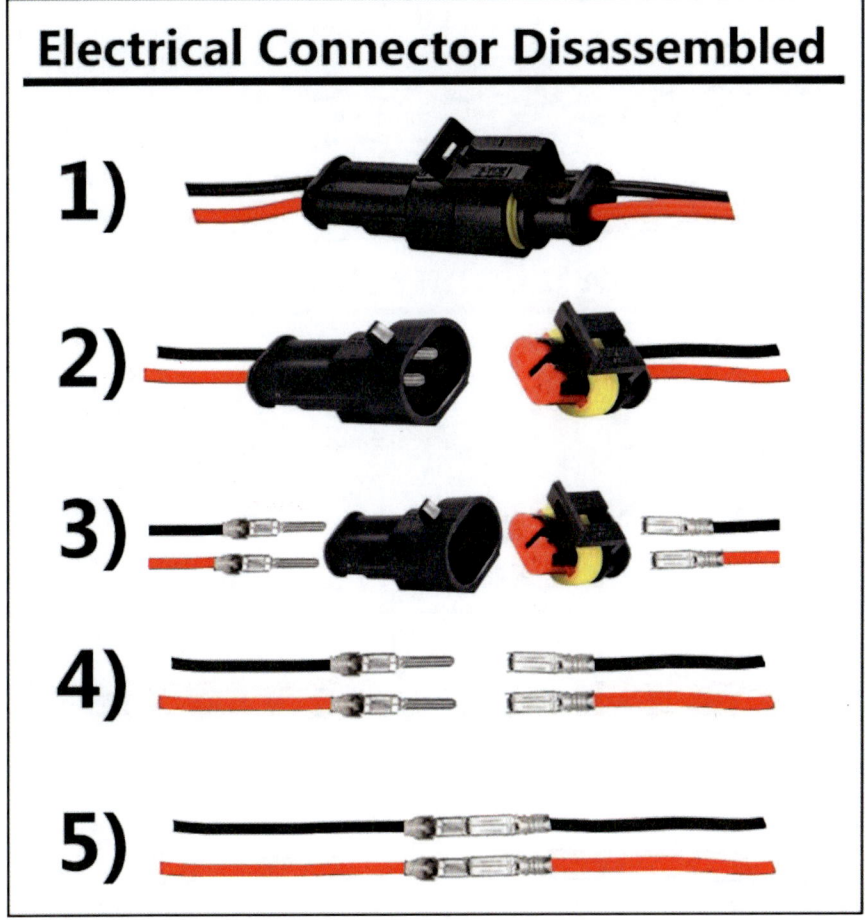

(In this illustration we see what is really inside an electrical connector once the plastic housing is removed. We have stripped the connector down step by step until all that remains is the wiring and the actual terminals that create the electrical connection. We need to understand how the connector's metal terminals connect as they are one of the places where an intermittent problem is likely to happen.)

The metal male terminal slides into the metal female terminal to complete the electrical connection and allow the flow of electricity. The **ONLY** thing holding the connection together is a spring-like tab on the female terminal. If this tab were to become worn, bent or broken, then a loose connection can occur and an intermittent electrical problem is likely to happen...

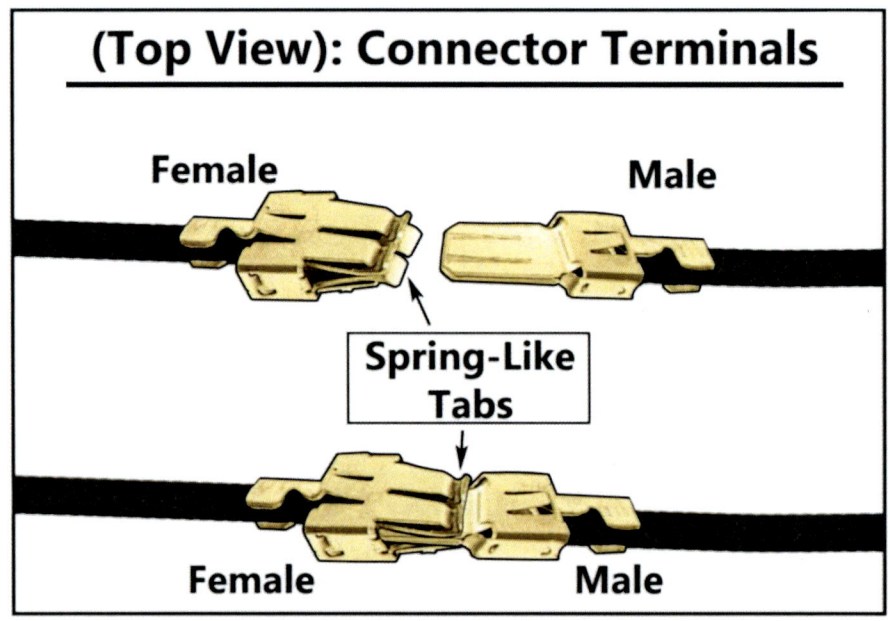

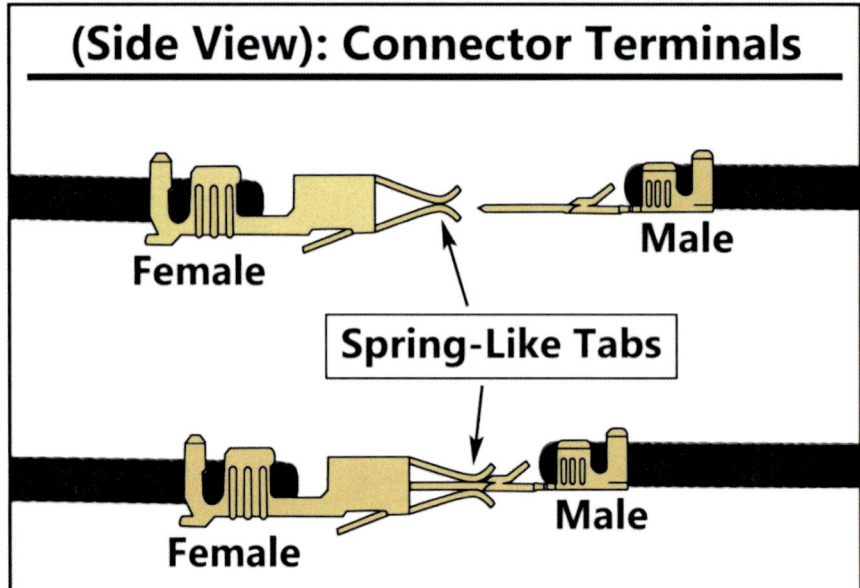

(In this illustration we see an example of a good male to female connection. The spring-like tabs on the female terminal is what holds the connection from separating once they are connected.)

Intermittent problems inside connectors usually occur because these spring-like tabs on the female terminal either become bent or broken due to years of age. Let's see now how an **intermittent** is created from bent female terminals inside a connector...

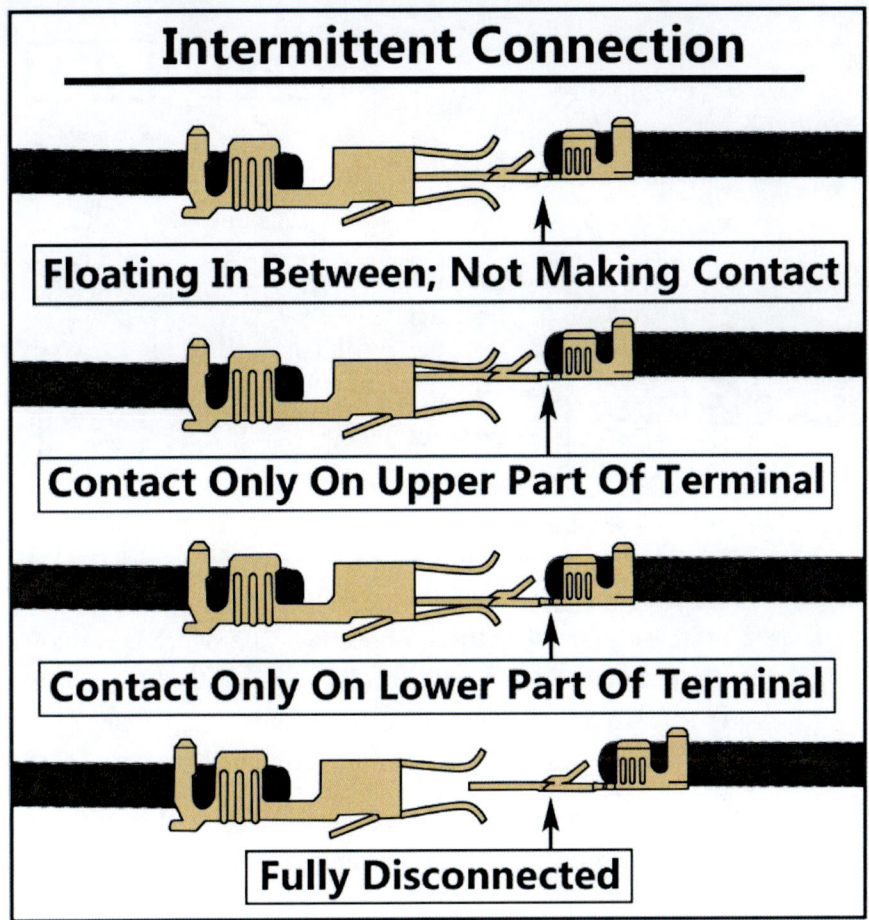

(Here we see examples of the metal terminals from our electrical connector now with the spring-like tabs bent out of place. Inside the connector, the terminals can take different positions that can cause the circuit to either be ON or OFF. Due to excessive heat, vibrations or movement near the connector, the metal terminals could easily fall out of contact and then reestablish contact again suddenly. This is how an **intermittent electrical problem** can occur inside connectors.)

This movement of the terminals happens within the connector and will not be able to be seen. What WILL be seen is the circuit turning ON and OFF randomly because of bad contact occurring within the connector. The terminals are constantly making or breaking contact therefore creating an electrical problem that seems to "randomly" happen.

Intermittent Internal Switch Contact:

Another common cause of an intermittent electrical problem is at the electrical switch for the circuit. The switch also has contacts and terminals that can wear out or become bend out of place over time. This will also eventually cause a bad electrical contact within the switch. Lets take a look at the internals of a switch and see how an intermittent problem can be created in a switch…

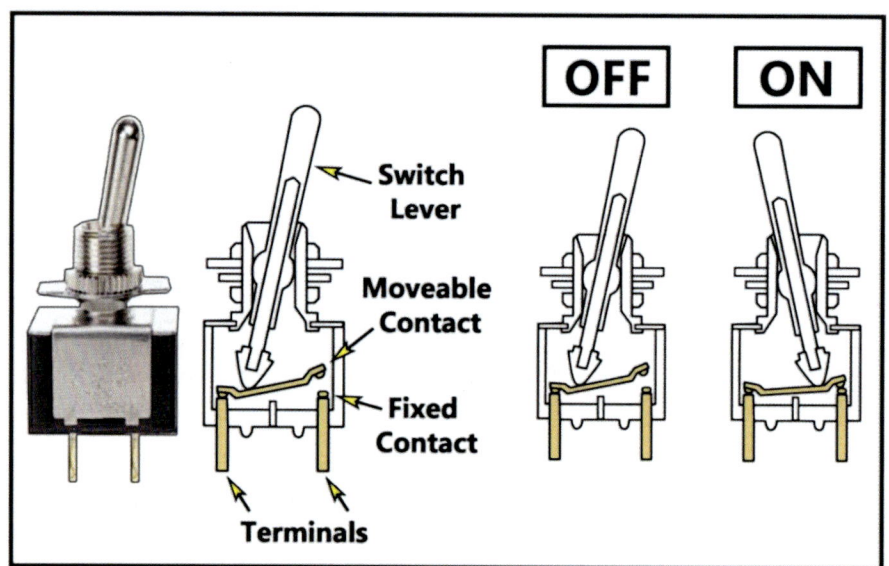

(In this illustration we notice the internals of an electrical switch. To the right we notice how the internals move to establish a good electrical connection. One switch is in the OFF position and the other is in the ON position. The moveable contact inside the switch moves with the movement of the switch lever to create or break electrical contact depending on position. The main part to focus on is the movable contacts inside the switch.)

If the moveable contacts in the switch were to become worn or bent, then the same intermittent effect as in the electrical connector can happen. Let's see now how this intermittent problem may occur..

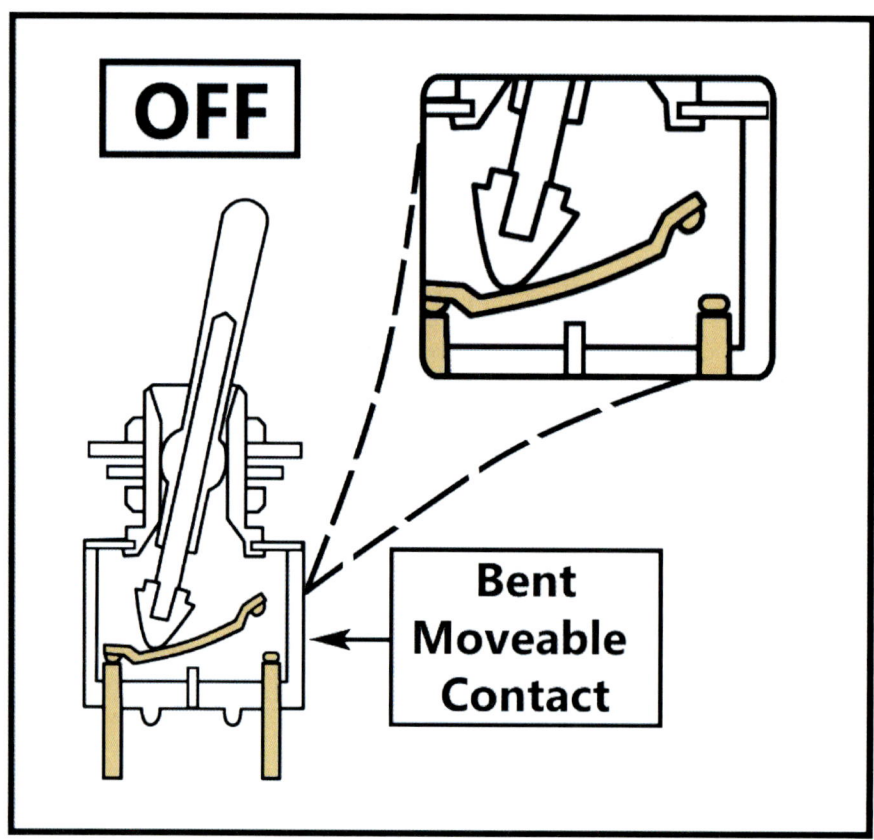

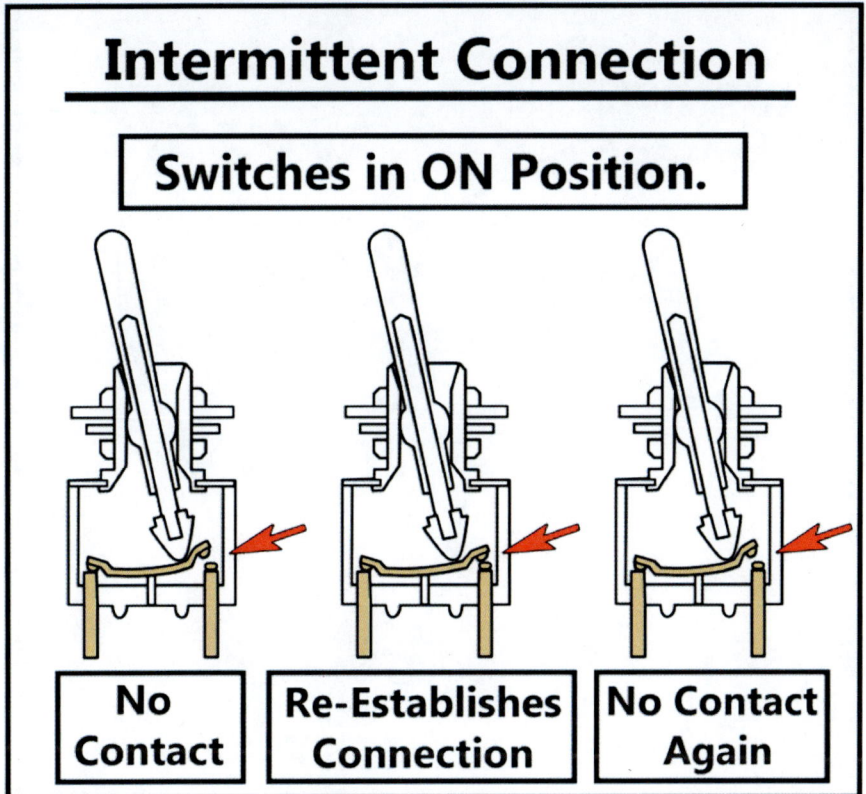

(In these illustrations, we see how a switch can be the cause of an intermittent issue. The moveable contact is bent and isn't making a solid electrical connection while in the ON position. Due to excessive heat or vibrations near the switch, the contact can easily reconnect and disconnect thereby creating an intermittent problem like in the electrical connector.)

This movement of the contacts happen within the switch and will not be able to be seen. Again, what WILL be seen is the circuit turning ON and OFF randomly because of a bad connection occurring within the switch. The contacts may unpredictably make or break electrical connection therefore creating an electrical problem that seems to "randomly" happen.

Intermittent Light Bulb Filament:
A third example of an intermittent problem can actually occur inside the light bulb itself. The internals may be worn or loose just like in the electrical switch.

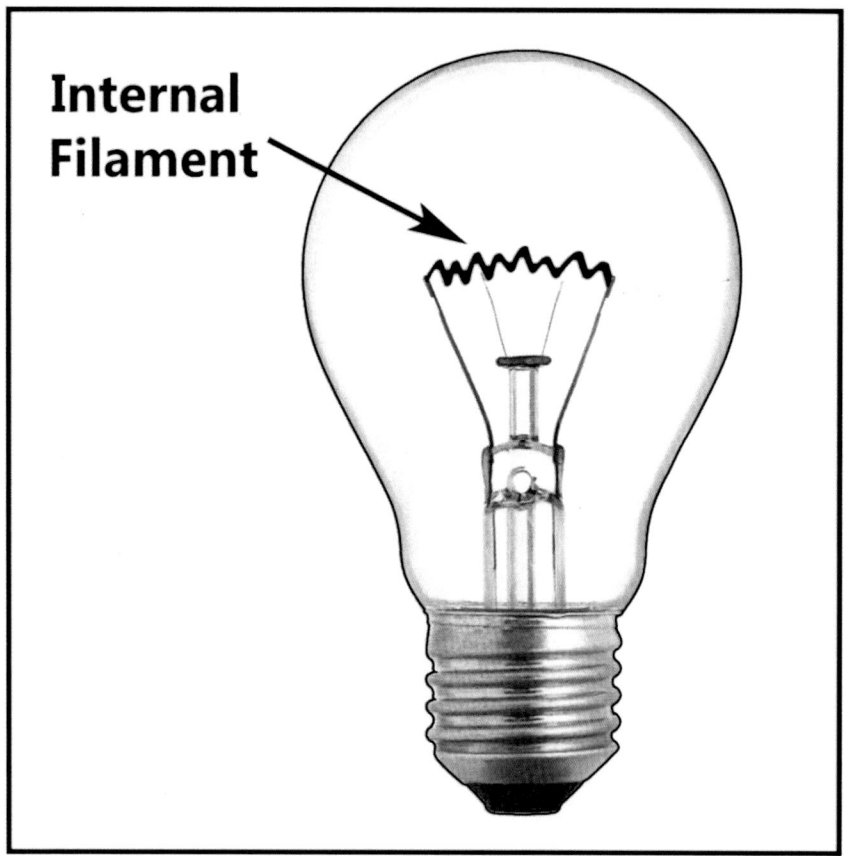

(This is what a good light bulb would look like. The light bulb's internal filament is secure and not broken or out of place from the conductive supporting bars.)

(In these illustrations we see how the electrical device itself can be the cause of an intermittent issue. The internals inside are partially broken and don't always make a solid electrical connection. Due to excessive heat or vibrations near the electrical device, the internals can easily disconnect and reconnect thereby creating an intermittent problem.)

Note: In this previous illustration we used a light bulb as an example of how an electrical device can be the cause of an intermittent problem. The electrical device doesn't necessarily have to be a light bulb. This can happen to any device regardless of it being a fan, a heater, a motor, etc. This example holds true for any electrical device because the internals can become broken and intermittently make or break contact similarly to filament in the light bulb.

There can be many areas in the circuit where an intermittent electrical problem may happen. A connector, a switch and the electrical device are only three examples of possible causes. The thing to really keep in mind is that wherever there are contacts and terminals in the circuit, there is a potential for creating intermittent effects. Also remember that the constant making and breaking of electrical connection is usually caused by either excessive movement or heat within the circuit. Let us now look at how intermittent problems can be categorized depending on what caused the contacts or terminals to move out of place and break the circuit intermittently...

Types of Intermittent Electrical Problems:

Most intermittent problems might create the same random ON or OFF effect but they may not be caused by the same thing. These problems can all generally fit into two main categories.

- **Vibration Related Intermittent:** Occurs when poorly connected parts of the circuit are moved out of place or whenever there are vibrations nearby the circuit to cause internal movement within an electric part. This excessive movement may cause loss of contact or loss of terminal connection inside a part in the circuit. This causes a temporary open circuit until vibrations again cause the internals of a part to move back into contact to where it was connected.
- **Temperature Related Intermittent:** Occurs only when the circuit is exposed either hot or cold. One example can be when the circuit has been ON for a long enough time and the temperature of the circuit has increased. This heat causes expansion of every internal metal parts of the circuit which then causes worn contacts and terminals to possibly move away from each other during expansion. The circuit may work fine for only 20 or 30 minutes and then suddenly cut OFF once the temperature rises to cause enough expansion of metal parts.

Depending on the type of intermittent electrical issue the testing methods for solving them would be different. For vibration related intermittents the better tool to use for the job is the graphing multimeter. For temperature related intermittents we would rely on other slightly different methods for diagnosis. One thing that remains constant between the two types is the need to somehow reproduce the intermittent while we are performing our testing. I will elaborate on this in the following chapters...

Ch.3: Vibration-Related Intermittent Issues:

In these types of problems we will notice that the circuit may cut ON and OFF quite often. Usually this kind of intermittent will continue to happen regardless of how long the circuit has been ON or how hot or cold it is around the circuit. When an intermittent problem meets these given symptoms we can assume that it is likely vibration-related in nature.

The Graphing Multimeter:

The test method I will provide for finding this type of intermittent problem is intended to provide a very simple methodical plan to finding the source of the problem. The tool of choice for this type of job will be the **graphing multimeter**. You will understand why this is the better tool for solving intermittent problems after understanding the exact nature of the problem.

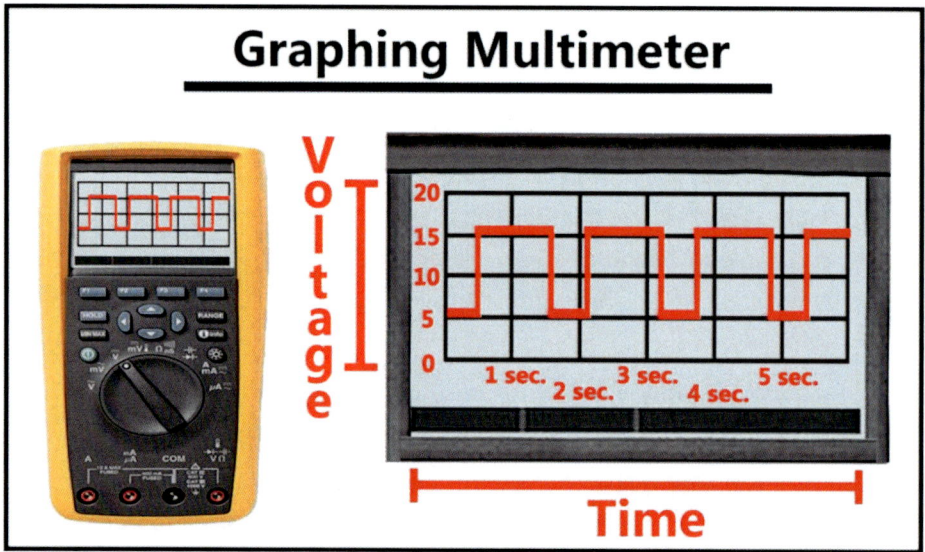

Note: The graphing multimeter is used the exact same way as you would use the normal multimeter except now the meter has the option of using the GRAPH function to display the measurements in a graph rather than in number. The graph display has a faster reaction time to voltage changes. The graph meter will be able to catch sudden voltage changes that may indicate a problem in that part of the circuit.

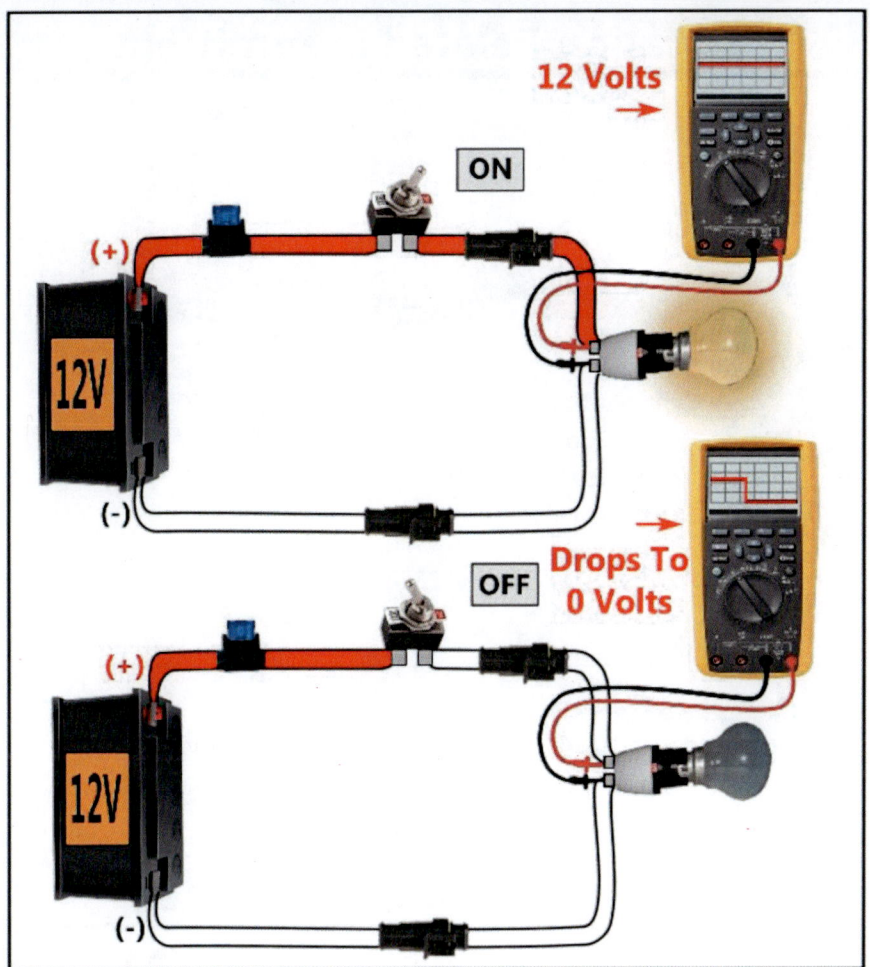

(Example of how the graphing multimeter graphs voltage. The graphing multimeter has been set to DC volts and the probes placed at the electrical device for monitoring. In the upper circuit, the constant 12 volt reading from the battery is displayed as a horizontal line on a part of the vertical scale that matches the amount of voltage being read on the meter. In the second circuit, we turn the switch off to see how the graphing multimeter displays this event.)

The graphing multimeter can display voltage changes over a period of time which is very useful when tracking down intermittent problems. It is also faster at reacting to voltage changes that may last only a fraction of a second that you might miss if you were to use a normal multimeter. Let us take a look at another examples of how the graphing meter reacts to voltage changes now in a vibration related intermittent...

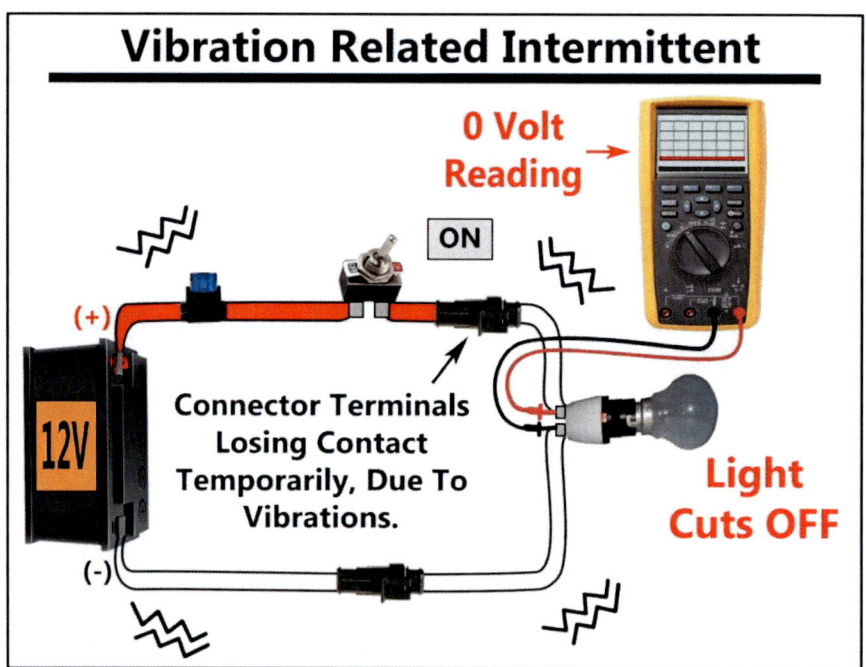

(In this version of our circuit we have a problem with a vibration related intermittent issue. We notice that the problem starts whenever there is movement or vibration near the circuit. When the circuit is exposed to vibrations or movement, the light turns OFF. While the circuit is OFF the graphing meter reads a 0v flat line at the electrical device.

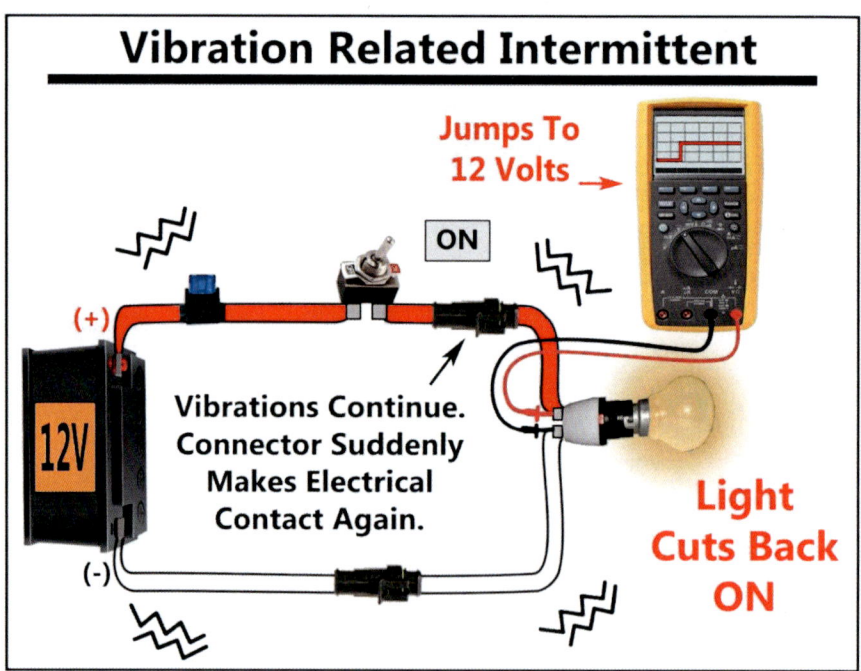

(As vibrations continue the circuit may suddenly cut back ON and the meter immediately picks up the voltage change and the reading jumps up to the 12v level on the graph.)

In this example, the problem is inside the power side connector that had worn out terminals, causing intermittent electrical contact. The connector is able to disconnect and reconnect randomly depending on the movement or vibrations near the wiring. This causes a seemingly randomly ON and OFF effect in the circuit.

Notice how the connector doesn't look damaged or worn at all from the outside even though it is the cause of the intermittent issue. This is commonly the case for most electrical parts you will come to test. The damage is internal and no visual signs of damage may be noticed.

Commonly the movements to the circuit may be created by things such as building vibrations near the wiring for the circuit, or maybe rats scurrying through the walls of where the circuit is installed or even engine vibrations if the circuit belonged to a car or other motored vehicle. Regardless of the source of the vibrations, we can diagnose it using our graphing meter to pinpoint exactly where in the circuit the problem is.

Reproducing the Intermittent For Diagnostic Purposes:

When solving any type of intermittent issue we will have to attempt to reproduce the intermittent while at the same time testing for voltage changes with our graphing meter. This can easily be reproduced for a vibration-related intermittent by simply wiggling the wiring and tapping on other parts of the circuit while monitoring the meter readings.

The Wiggle and Tap Test:

The wiggle and tap test used for vibration-related intermittents is fairly easy to follow. We simply need to shake the wiring or tap on all electrical parts of the circuit including all **connectors, fuse terminals, relay and relay terminals** (if equipped), **electrical devices and terminals, power source connections,** etc. Essentially, anywhere where there is an electrical connection for the circuit can be a possible cause for an intermittent.

Diagnosing Vibration-Related Intermittents:

In the following we will go through some example problems and focus on how to solve vibration-related intermittents from what we have learned so far..

Intermittent Example #1:

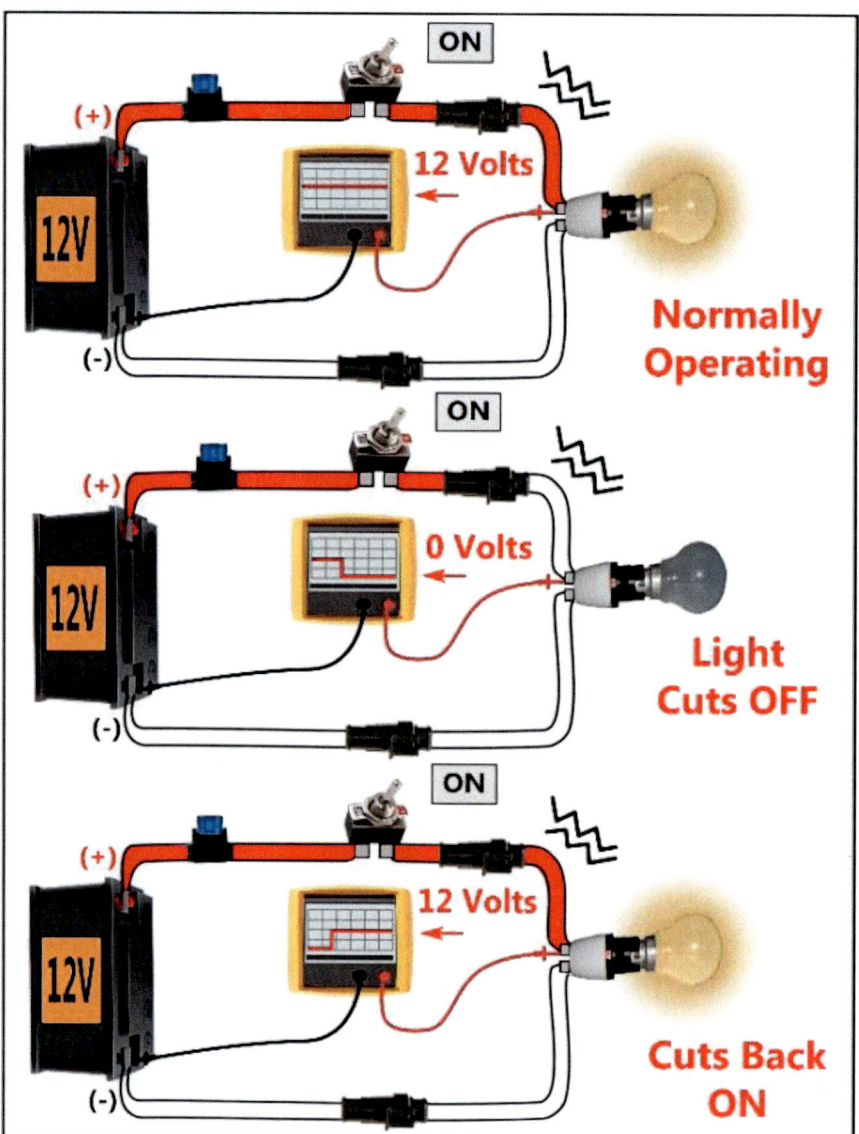

(We start by testing at the electrical device, one probe to the battery negative one probe to the positive side of the light bulb. We attempt to reproduce the intermittent problem by shaking the wire near the light and notice that the light begins to cut OFF intermittently. The voltage reading on the graphing meter drops out to 0v whenever the intermittent is in effect. The voltage reading jumps back up to 12v whenever the light cuts back ON.)

Important Note: The dropout in voltage seen on the graphing meter may only last for a split second. This is why a graphing meter is the better tool for this sort job since we will be able to see all the dropouts in voltage as they happen. A graphing meter allows us to see the intermittent voltage changes better because it graphs the voltage changes that are happening over a period of time (usually graphs 5 seconds worth of voltage readings).

Objective During Testing: The goal for this diagnosis is to find an area in the circuit where the voltage no longer drops out during the intermittent problem. The circuit may still intermittently be turning ON and OFF but the dropping out on the meter will stop at a certain test point. This is how we will know that we are very close to finding the cause of the problem.

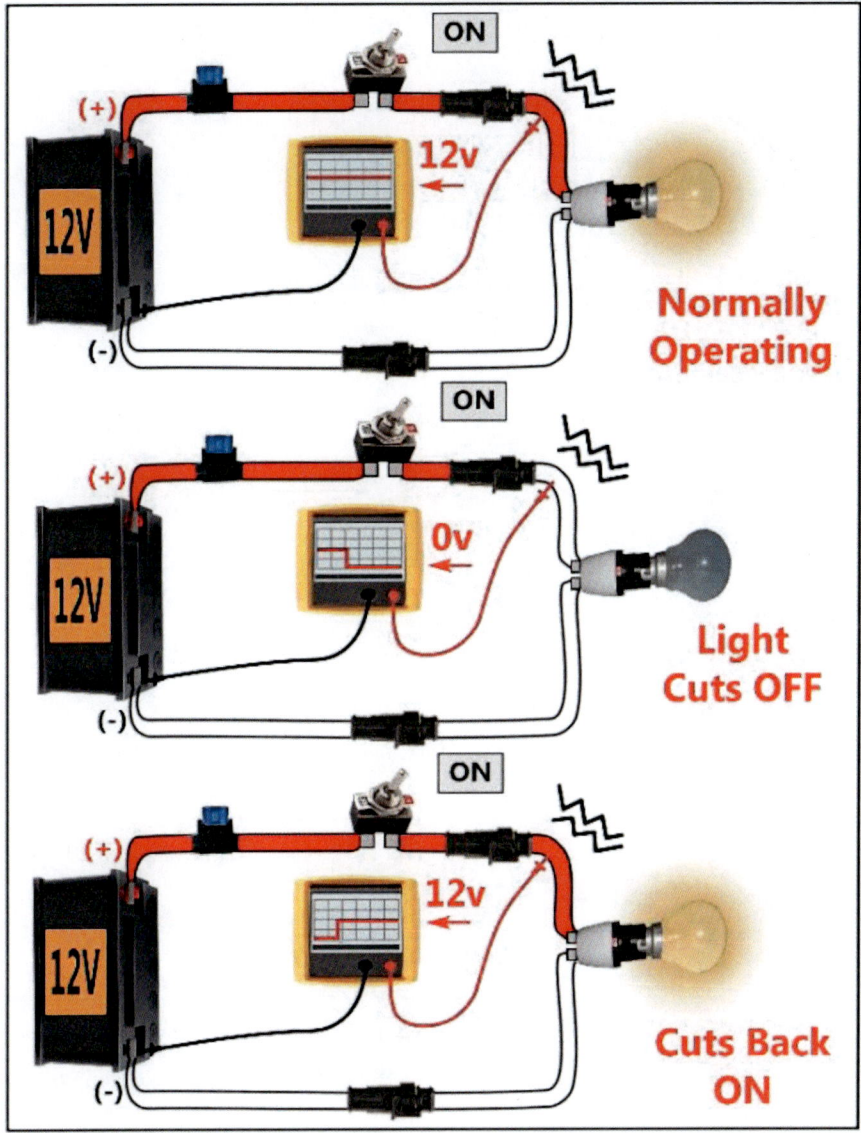

(In this illustration we continued our voltage testing of the power wire, with one probe to ground and the other probe testing at a different part of the circuit. We have moved our red probe to one side of the connector on the power wire. The meter is STILL dropping off in voltage as we reproduce the intermittent by shake the wiring. We must continue diagnosis by testing at another point in the power wire to find the cause.)

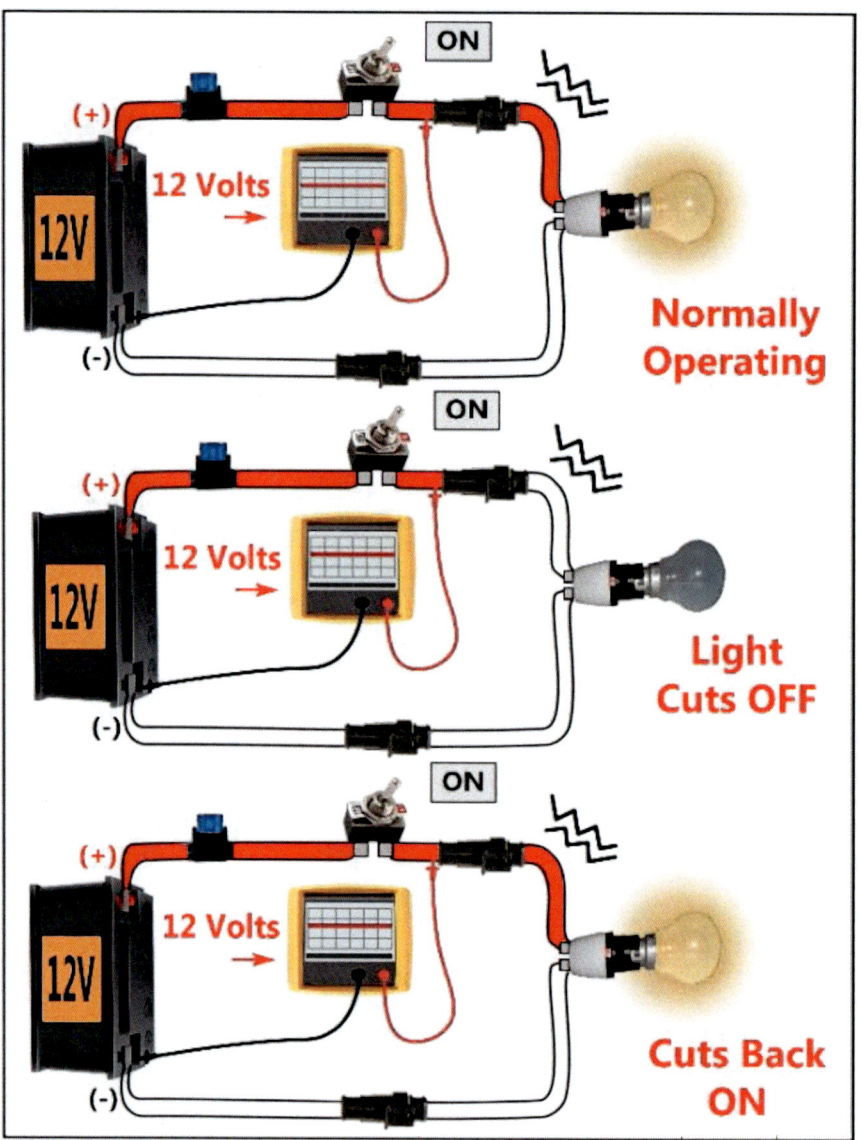

(In this illustration we continue our voltage testing by moving our red probe now to the other side of the connector on the power wire. The meter STOPPED dropping off in voltage as we shook the wiring near the connector.)

This reading on the graphing meter indicates that we are very close to finding the problem even if we can't see it right away. The problem is somewhere between where the red probe is now and where the red probe was installed previously. The only thing between those two test points is the connector. Upon further inspection of the connector, we noticed that the spring-like female terminals inside it had been bent outward causing an intermittent contact with the male terminal.

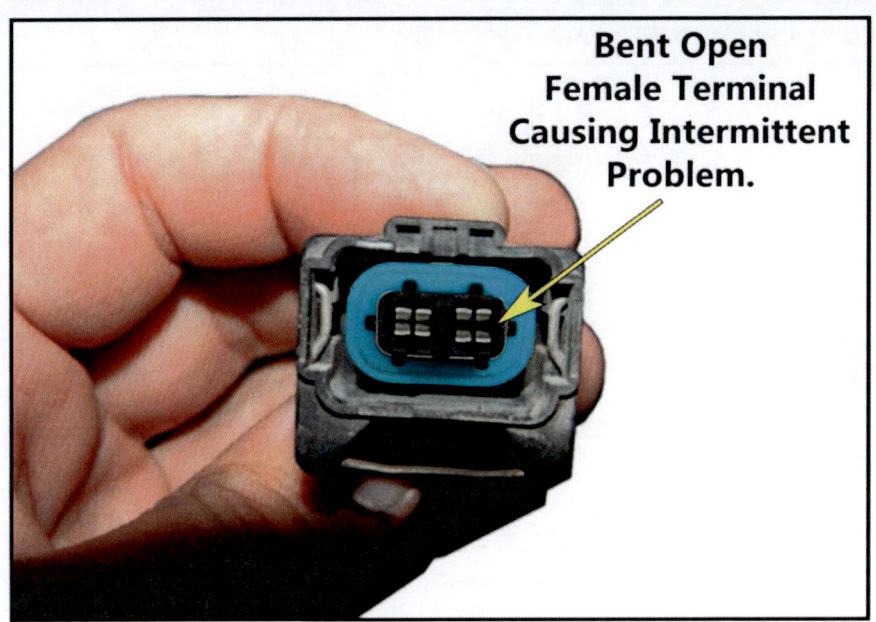

(This image shows the inside of a two wire connector that caused an intermittent electrical problem. One of the terminals inside the female connector is bent open and caused intermittent contact with the male terminal, once it was installed.)

You may get away with repairing this sort of problem by bending the terminal back down to make a tighter connection, but in a more professional environment you may have to replace the damaged terminal instead.

The reason why replacement of the terminal is better is because the terminal may already be too brittle or worn out over the years and might create a problem once again along the line.

Now that we have seen how a connector can be one of the causes to an intermittent electrical issue, let's look at another example of a vibration-related intermittent and how we can solve it as well…

Intermittent Example #2:

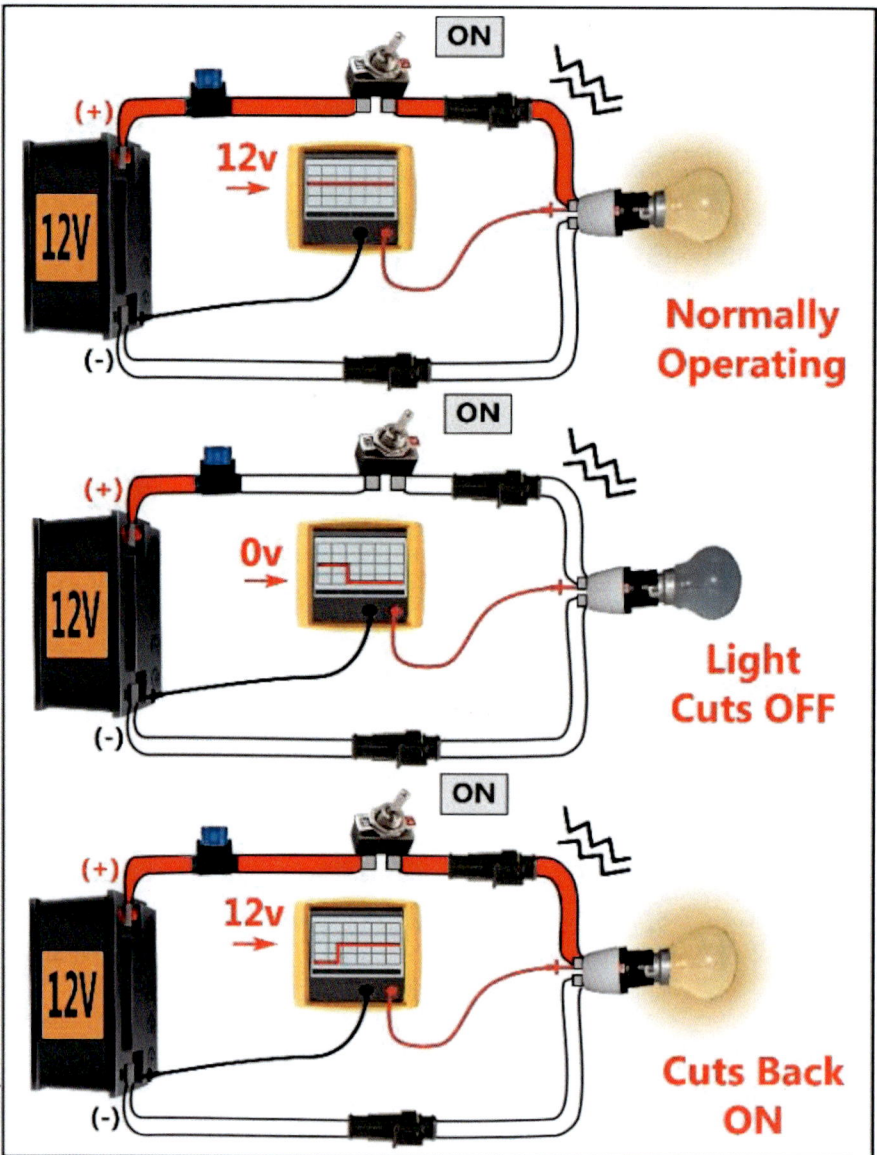

(In this version of our intermittent problem we have the same complaint as before. The circuit turns ON and OFF randomly on its own. We begin by testing at the electrical device, one probe to the battery negative, the other probe to the positive side of the light bulb. We start shaking all the wiring and notice the light begins to cut OFF intermittently. The voltage reading on the graphing meter drops out to 0v whenever the intermittent is being reproduced. The voltage reading on the graphing meter jumps back up to 12v whenever the light cuts back ON. We must continue testing at another point to find out the problem.)

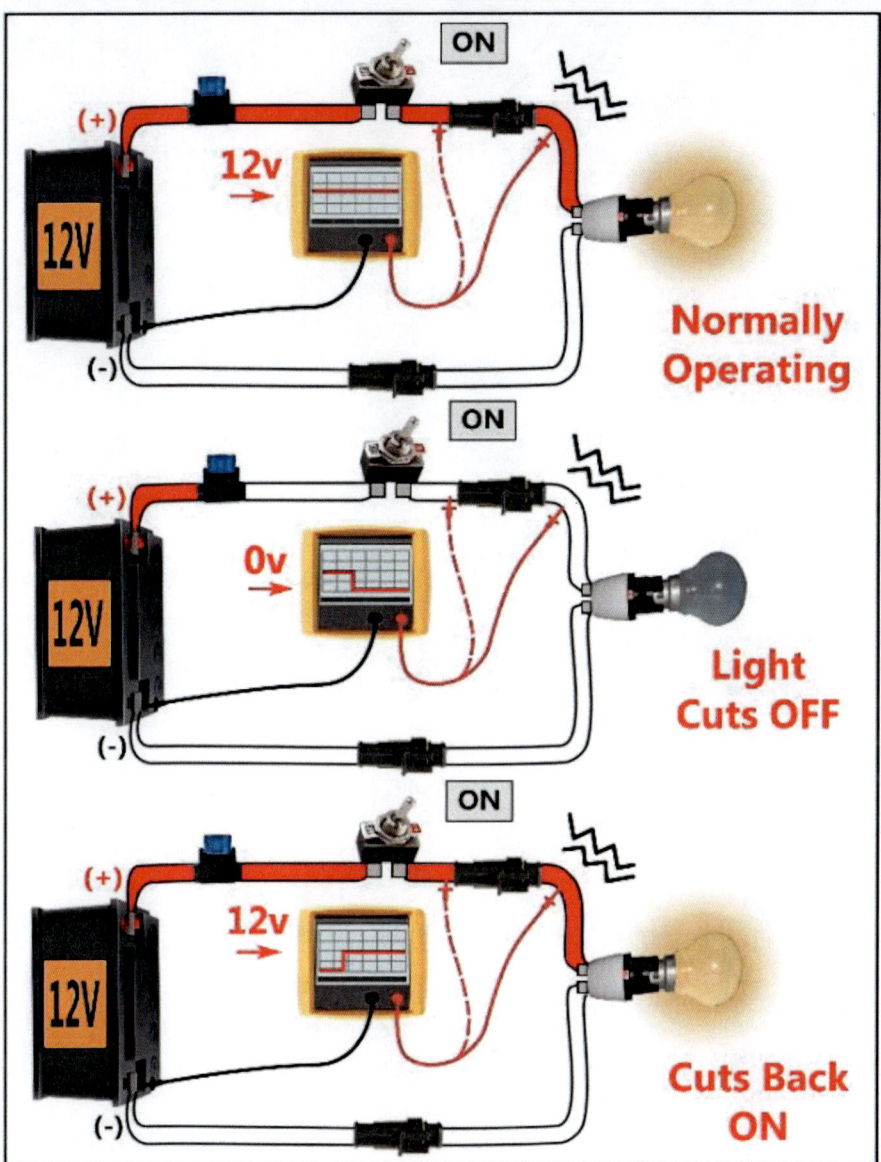

(In this illustration we have continued our voltage testing at another point in the circuit. We have tested both sides of the connector while shaking the wiring yet STILL the meter reading continues dropping off in voltage as we reproduced the intermittent. This indicates that the problem is still further back towards the battery.)

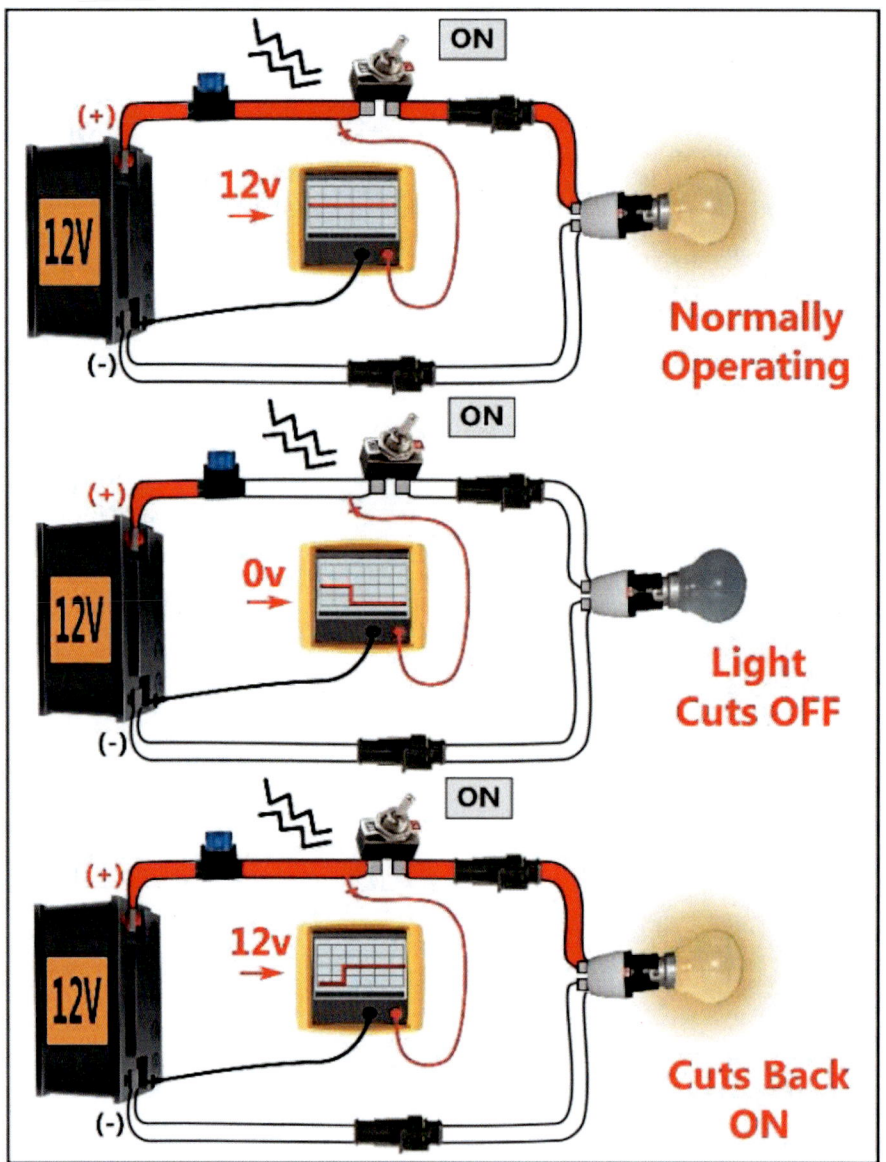

(In this image we continued voltage testing at the front side of the switch. STILL the meter reading continues dropping off in voltage as we shake the wiring. This indicates that the problem is STILL further back towards the battery. We must continue at another test point until the dropping out of the voltage reading stops.)

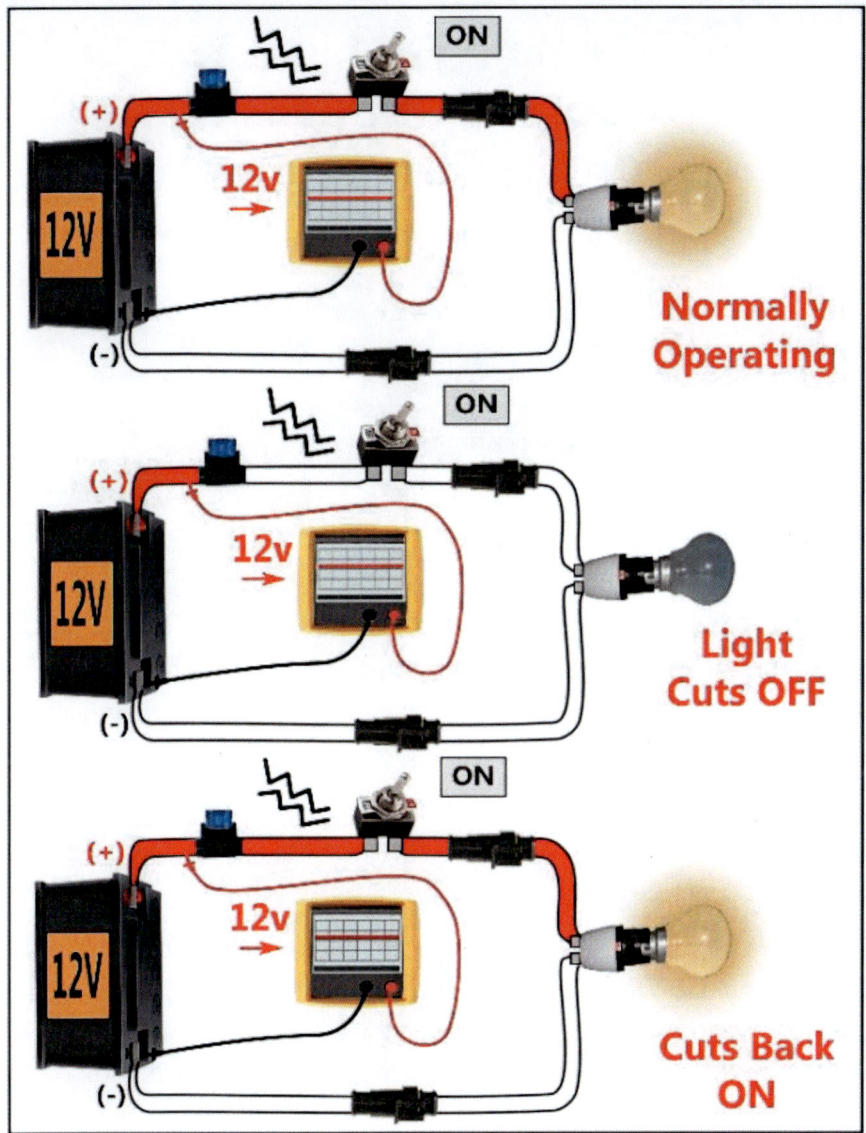

(In this illustration we continue by moving our red probe now to the front side of the fuse holder. Now the meter reading STOPPED dropping out in voltage as the intermittent turned the circuit ON and OFF.)

This reading indicates that we are very close to where the problem is. The problem is somewhere between where the red probe is now and where the red probe was installed before. The only thing between those two points is the fuse and fuse holder.

Upon further inspection of the area we noticed that whenever we tap or move the fuse holder, the light turns ON or OFF. When we removed the fuse to inspect the fuse holder terminals we noticed that one of the female terminals was bent out of place causing intermittent contact with the male terminal.

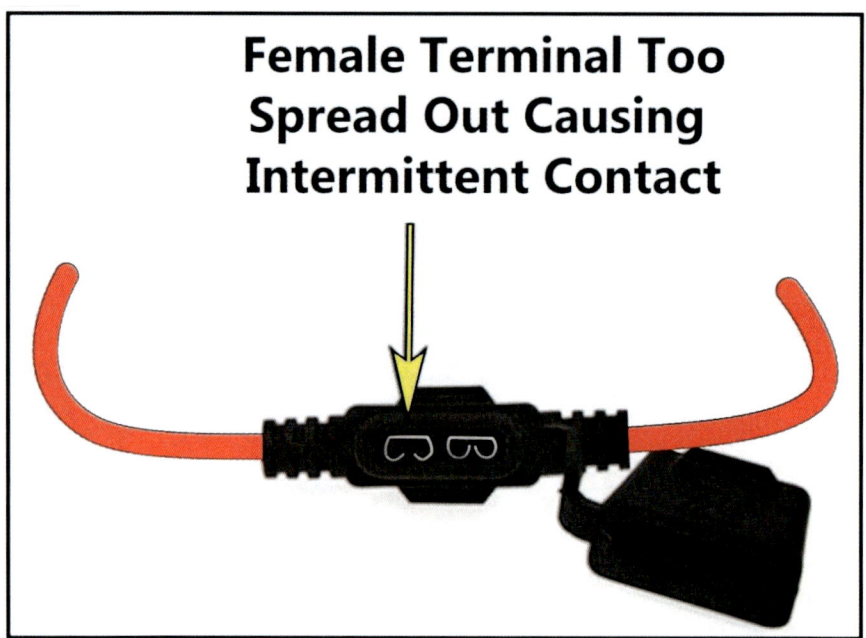

(Picture of the female terminal that was causing the intermittent ON and OFF issue in the circuit. One of the terminals inside the fuse holder was bent open and caused intermittent contact with the male terminal of the fuse.)

The repair may be as easy as bending the terminal back down to create a tighter fit for the fuse. In a professional environment, you would likely want to replace the fuse terminal so it doesn't create another problem at some point.

Notice how the bent terminal inside the fuse holder seems only barely bent outward compared to the good one. This very small amount of bending is more than enough to cause an intermittent problem. Many times the problem is not as obvious as we would expect. Never overlook these minor things when testing out on the field. No electrical connection should be bend outward or loose.

Important Note: The reason as to how the fuse holder terminal became bent was apparent after realizing that the fuse had been previously replaced. During installation, the previous technician may have accidentally bent the terminal while installing the fuse.

Always look for these sort of clues as they can save you time on diagnosis with a simple inspection of a previously worked on area.

Let's now move on to a different example that may create a bit more of confusion when diagnosing.

Intermittent Example #3:

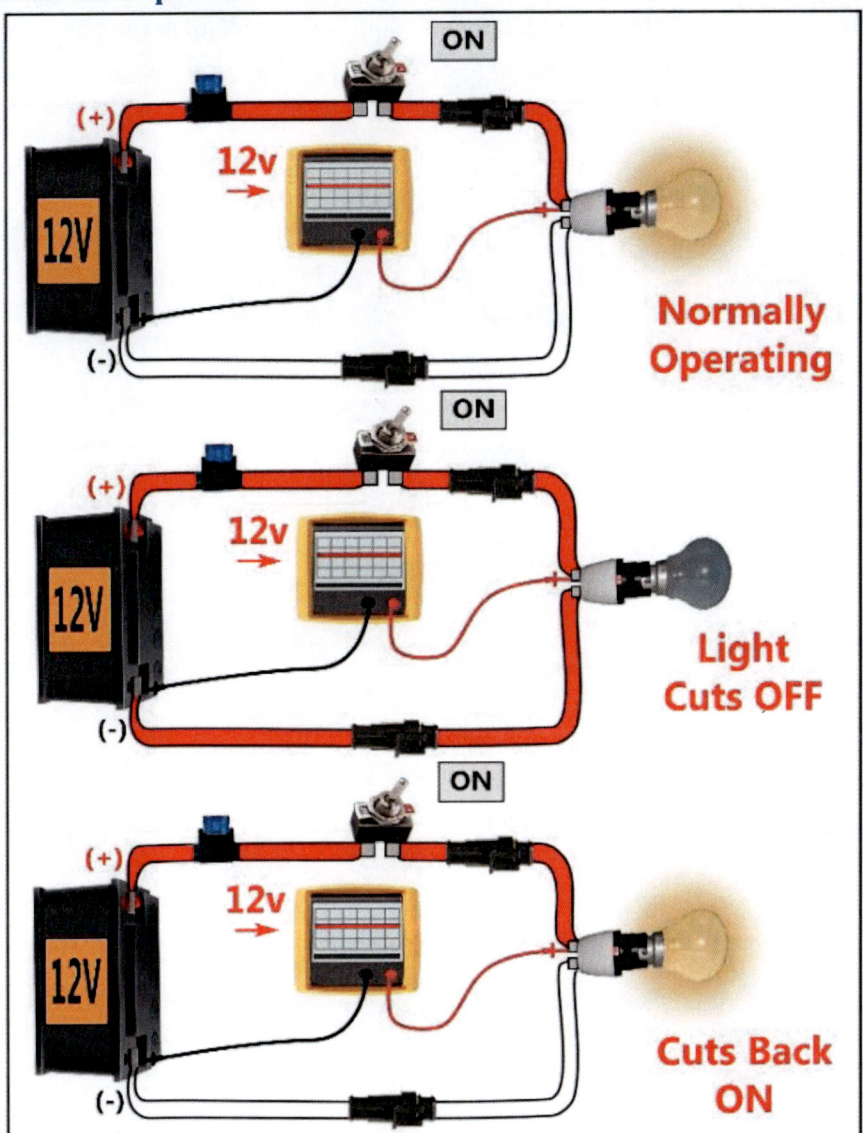

(In this version of our intermittent problem we have the same symptoms and complaint. The circuit turns ON and OFF randomly as in the circuits before. We begin testing at the electrical device, one probe to the battery negative, one probe to the positive side of the light bulb. We shake all the wiring in the circuit and notice the light is cutting OFF intermittently. The graphing meter displays 12v throughout the entire time the intermittent is in effect.

The main difference between this example and the other two examples is that the voltage reading on the graphing meter stays at 12v regardless of the light cutting ON or OFF. Right away there are three possibilities to how this can happen. One, the electrical device (light bulb) itself is the cause of the intermittent issue. Two, the connector terminals or socket where the light bulb connects are loose or worn out. And Three, the problem is somewhere after the light bulb on the negative side of the circuit.

There are three possible culprits to this intermittent according to the meter reading taken. It can be either the light bulb, the light bulb connections, or a problem in the negative side of the circuit. Let's continue meter testing to see how we can narrow down on the cause.

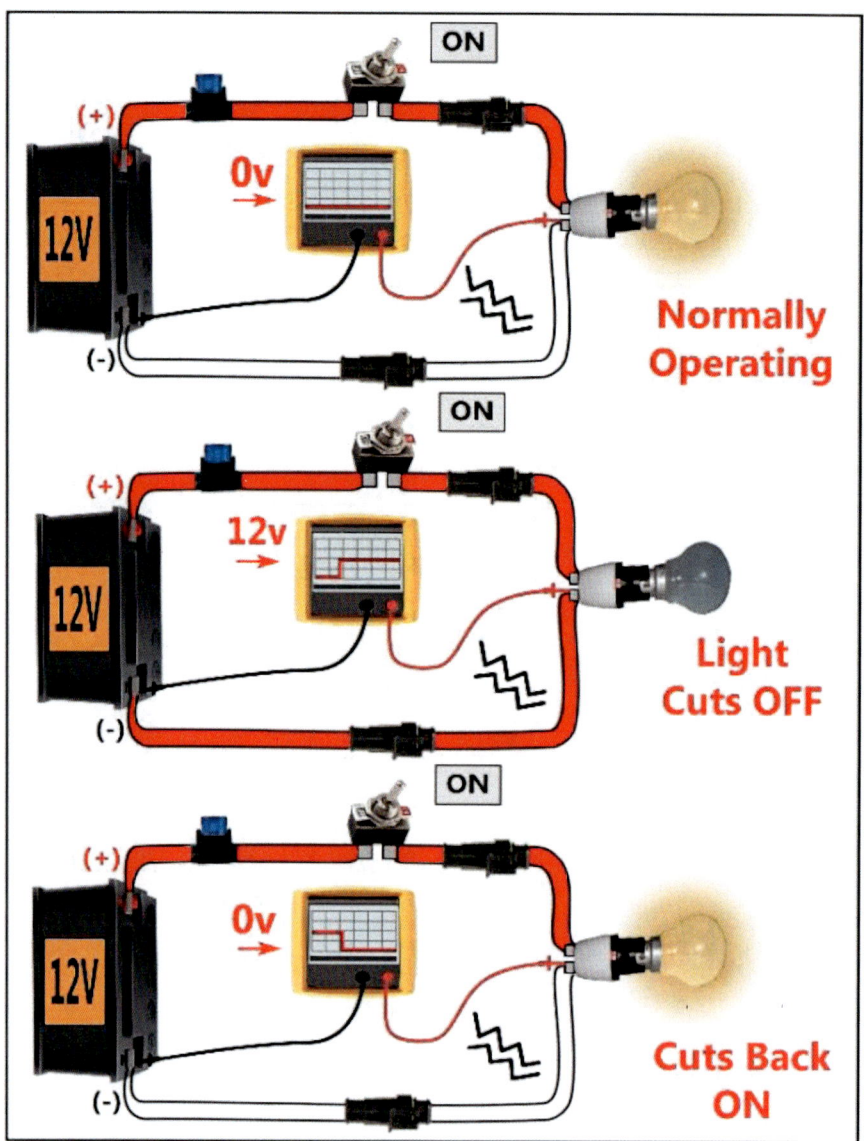

(In this illustration, we've continued voltage testing by moving our red probe now to the other light bulb terminal that connects to the ground wire. We begin by shaking the wiring and components to reproduce the intermittent problem. When the circuit is ON there is a 0v reading on the ground side (normal). When the intermittent problem cuts the circuit OFF the meter reading immediately jumps up to 12v (bad). This 12v reading indicates that the problem is not the light bulb or its connections. The 12v reading indicates that the problem must be somewhere ahead on the negative side.)

When the circuit turns back ON, the meter reading drops back down to the normal 0v reading. So far, we are on the right path but must continue testing at another test point on the negative wire until we find a test point where the spiking up in the voltage reading stops.

In this example, where the problem is on the negative side, the meter readings we are looking for are reversed. We start with a reading of 0v on the negative side when the circuit is working normally. When we measure a voltage spike up to 12v on the negative side, this means that there is an intermittent break somewhere ahead in the negative part of the circuit. The goal we are after as far as readings is to find a test point in the circuit where the meter reading stops jumping up from 0v to 12v when the intermittent is in effect.

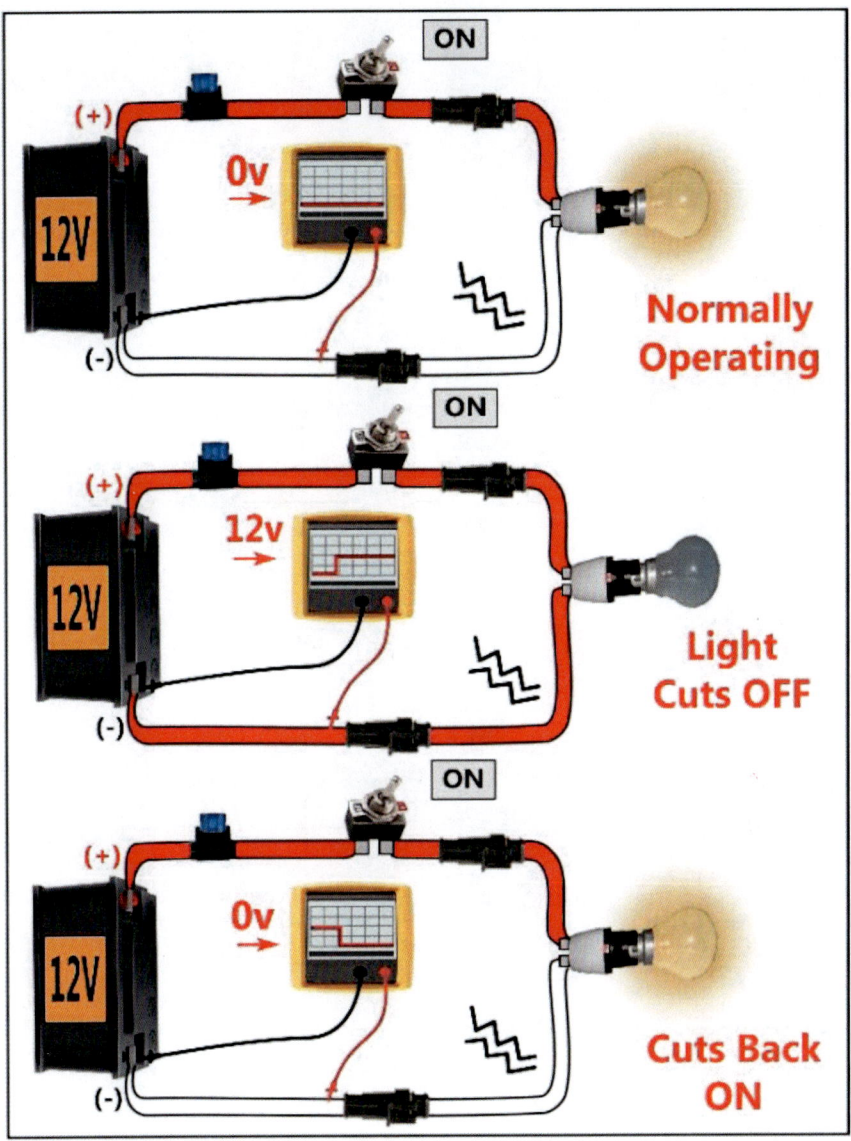

(In this image we continue our testing by moving the red probe to the negative side connector. We tested each side of the connector and the results were the same on either side. When the circuit is ON there was a 0 volt reading on the negative side. When the intermittent cuts the circuit OFF, the meter reading jumped up to 12 volts. Then when the circuit turns back ON, the meter drops back down to the normal 0 volt reading. These results tell us that the problem is still somewhere ahead in the rest of the negative side circuit.)

Since the voltage reading hasn't stop spiking up during the intermittent, we are still not near the source of the problem. So far based on the test results we have gathered, we know that everything up to the negative side connector is good. The only thing left to test now is the rest of the wiring from the negative side connector to the battery terminal on the negative side of the battery. Let us continue our testing now and see what else we find out about this intermittent problem..

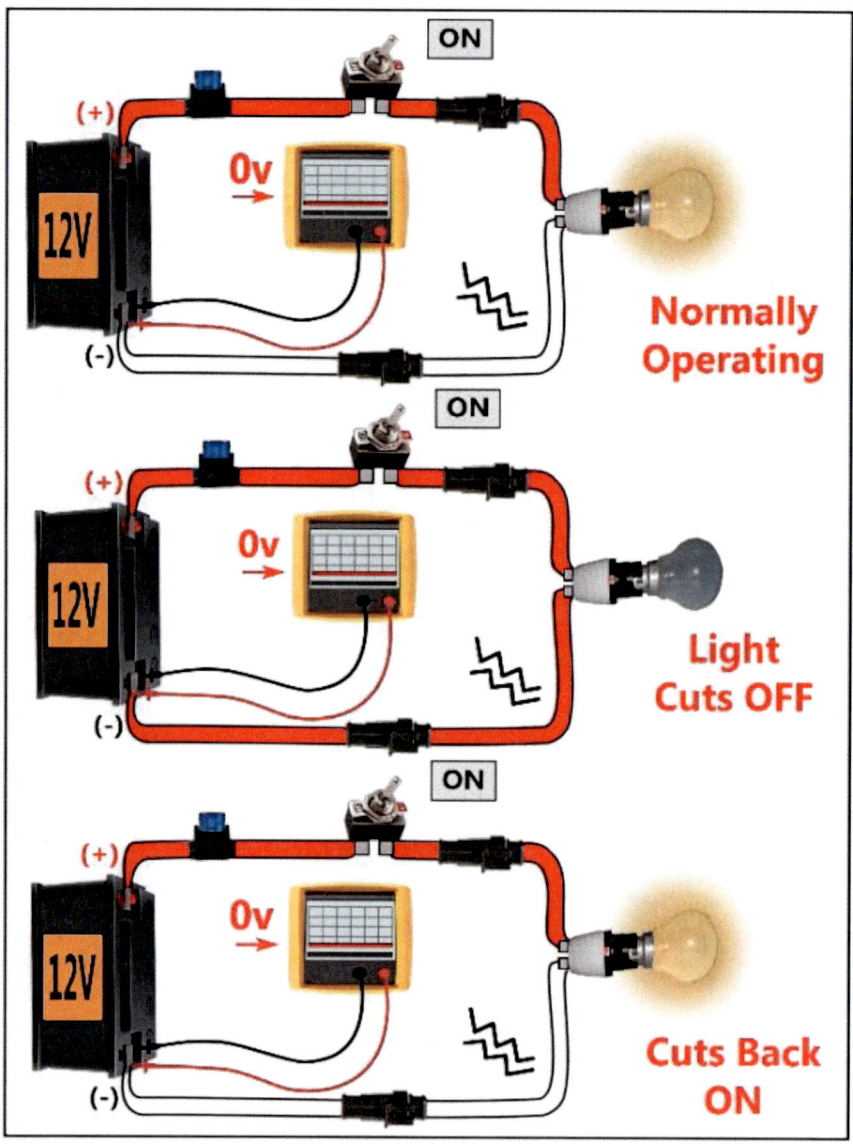

(We continue our testing by moving our red probe to the negative cable clamp. The black probe is on the negative terminal and the red is on the negative clamp. We are now testing for a drop across the negative battery terminal and the negative battery clamp. The results were different now at this test point. The voltage stopped spiking up to 12v when the intermittent problem was active. By this result we can conclude that the problem must be between the most recent test point and the previous test point. The only thing between the battery clamp and the negative side connector is the wiring that connects the two.

Upon further inspection we noticed that the intermittent happens more often when we moved the negative battery cable. The reason for this was that the end of the negative battery cable was loose inside the cable's terminal clamp. The movement of the cable within the cable's terminal clamp was the source of intermittent electrical contact.

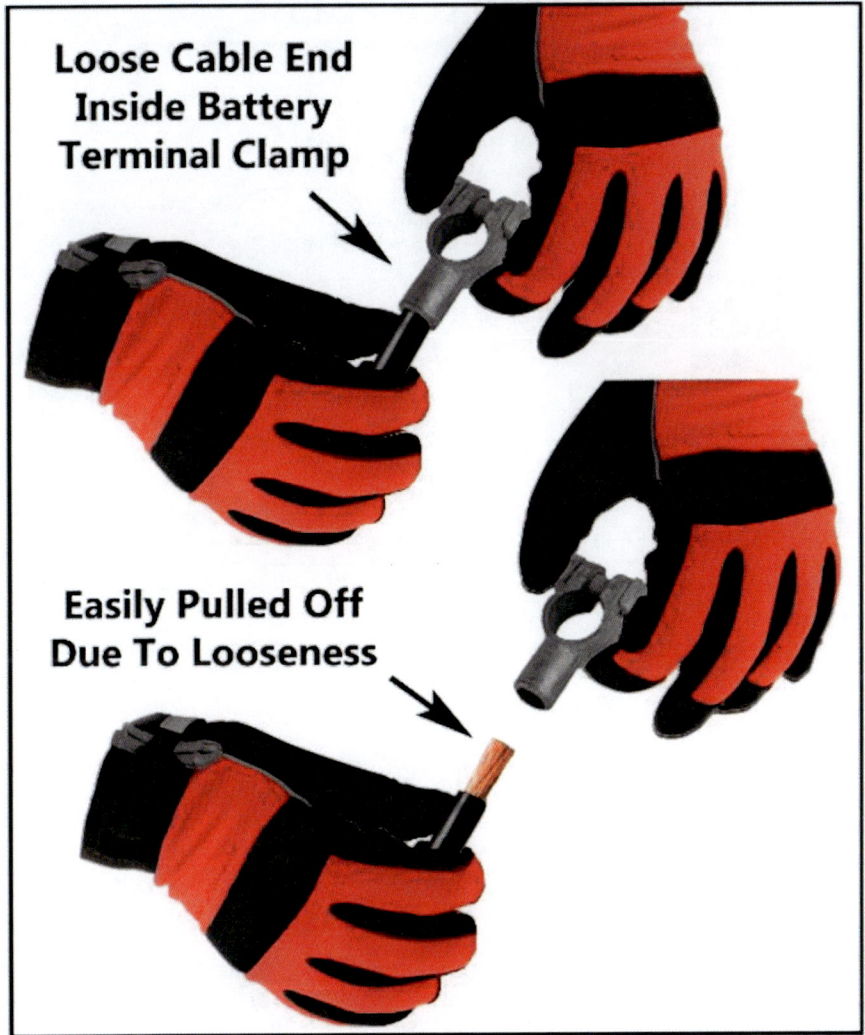

(This is image shows the source of the intermittent problem in our last example. The negative battery cable had become loose inside the cable terminal clamp causing intermittent electrical contact. The cable end was so loose that it was easily pulled off the terminal clamp.)

The terminal had a bad cable to terminal connection due to an improper terminal crimping job. The repair in this case requires that the terminal clamp be replaced with a new one and re-crimped or soldered professionally to create a good connection.

Important Notes About Testing Intermittents:

Note 1: If the circuit suddenly stays OFF permanently during intermittent testing, the next step would be to resort back to open circuit testing. We would have to voltage test across all parts of the circuit until we can find a point where the battery voltage is present. This is fully outlined in the previous book "How to Test Circuits Like a Pro: Part 1".

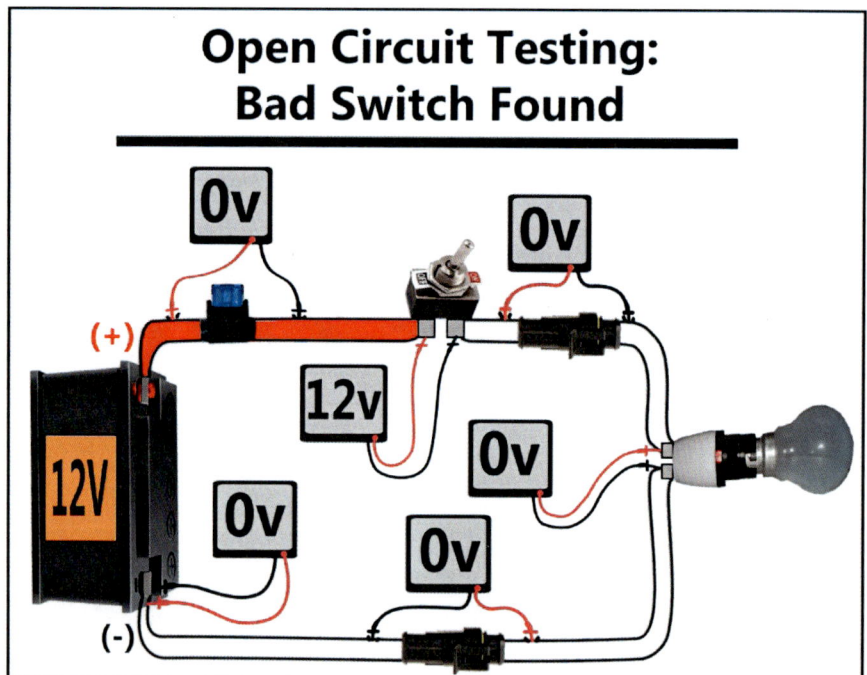

(Example of Open Circuit testing after the circuit stayed OFF during intermittent testing. The result indicated that the switch was the source of this problem. Make sure the switch terminals are not loose before replacing the switch.)

Note 2: If the intermittent problem cannot be reproduced for testing then we cannot diagnose the intermittent. By the nature of the problem, intermittents are known to be random. If we cannot reproduce the problem by shaking one part of the circuit, we must try shaking other parts until the problem CAN be reproduced. This includes all wiring, battery clamps, switch and switch terminals, light and light terminals, connectors, and fuse holders. If the problem still cannot be reproduced, then the problem may be temperature-related.

By now you should have a good understanding as to how you can use your meter to determine the cause of a vibration-related intermittent. We have seen different examples and what it takes to diagnose each of them. You may have also noticed that the diagnostic process of each example is generally very similar. This is because of the fact that the method to find any vibration-related issue using your graphing meter is basically the same regardless of the cause of the problem. Remember that the cause of an intermittent can be anything. Never overlook the little things when diagnosing an intermittent.

Now let us look at the second type of intermittent problem and how differently we would diagnose them..

Ch.4: Temperature-Related Intermittent Issues:

In this type of intermittent, the problem may start only after the temperature has changed around the circuit or depending on how long the circuit has stayed ON. You will notice that there is no effect to the circuit operation when performing a **Wiggle Test** like done in the previous type of intermittents. Instead we might notice that the circuit cuts ON and OFF only when the circuit has been allowed to be ON for a certain amount of time. This type of problem is known as the temperature-related intermittent.

Usually when a circuit problem meets these sort of temperature or time related symptoms, we can assume that the electrical problem will be temperature related. Let's take a look at an example of how a circuit may perform during a temperature-related issue..

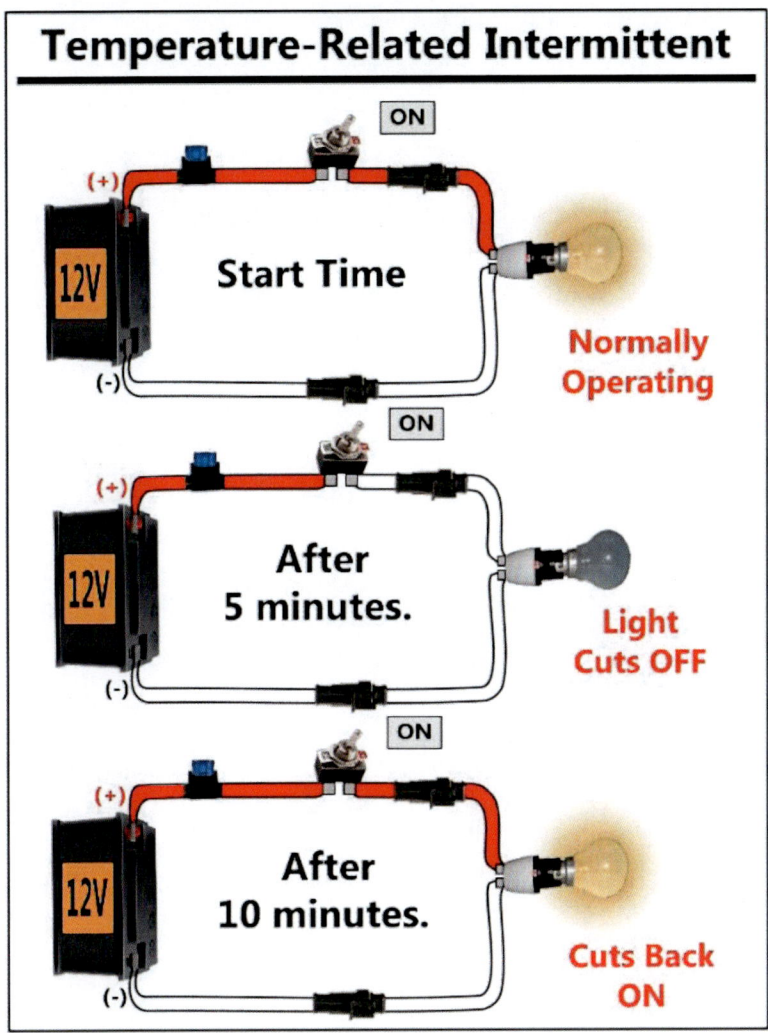

(In this example of a temperature-related intermittent, we notice that the problem starts whenever the circuit has been ON for some time. Whenever the circuit has warmed up, the light cuts OFF temporarily. Usually this CUT OFF happens within 5 minutes after the circuit has been turned ON. Then after about 10 minutes of waiting, the circuit will cool back down and the light turns ON again.)

The problem here is that this circuit only works for a short time up until it begins to get hot. When the circuit gets hot, the light turns OFF. When the circuit cools back down, the light will turn back ON again. Let's take a look as to why is this may be happening and how we can diagnose the problem effectively.

Methods For Diagnosing:

Before diagnosing a temperature-related intermittent, we must determine the exact symptoms of the problem. Depending on the symptoms we will use a different test method. There are three general ways to approach this type of intermittent. There is the Open Circuit Approach, the Graphing Meter Approach and then there is the more direct Temperature Changing Approach. We can choose to use either one of the methods depending on the specific conditions of the intermittent problem or personal testing preferences.

Lets take a look at an example of each testing approach using various versions of a temperature intermittent.

Open Circuit Approach: In one version of a temperature-related intermittent, the circuit may take a very long time to cut OFF and then cut back ON after the cooling down. If the circuit has a long enough OFF time, after the intermittent has caused the circuit to turn off, then we would resort to basic open circuit testing to solve the problem. We would begin our open circuit testing as soon as the circuit cuts OFF.

Intermittent Example #1:

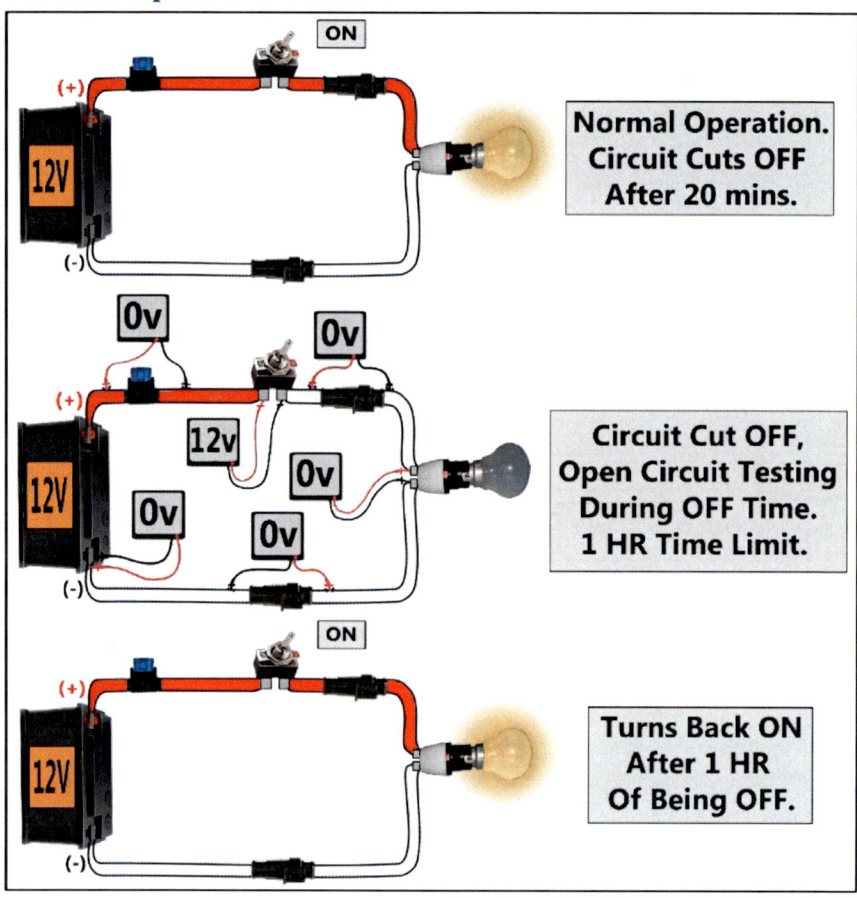

(In this example of the intermittent, we have a circuit that turns OFF only during warm weather. The circuit cuts OFF after about 20 minutes of use. The circuit turns back ON after being allowed to cool off for 1 hour. In this example we will approach the problem by simply using open circuit testing methods to find the problem during it's OFF time. We started by testing for voltage across various parts of the circuit to try to find the problem. We have a time limit of about 1 hour to find the problem before the circuit has cooled down enough to turn back ON. If we did not get the chance to find the problem during this OFF time, we must wait another 20mins for the next OFF time to continue open circuit testing. In this situation the cause of the problem seems to be the switch.)

Let's take a look at how the switch is causing this intermittent problem..

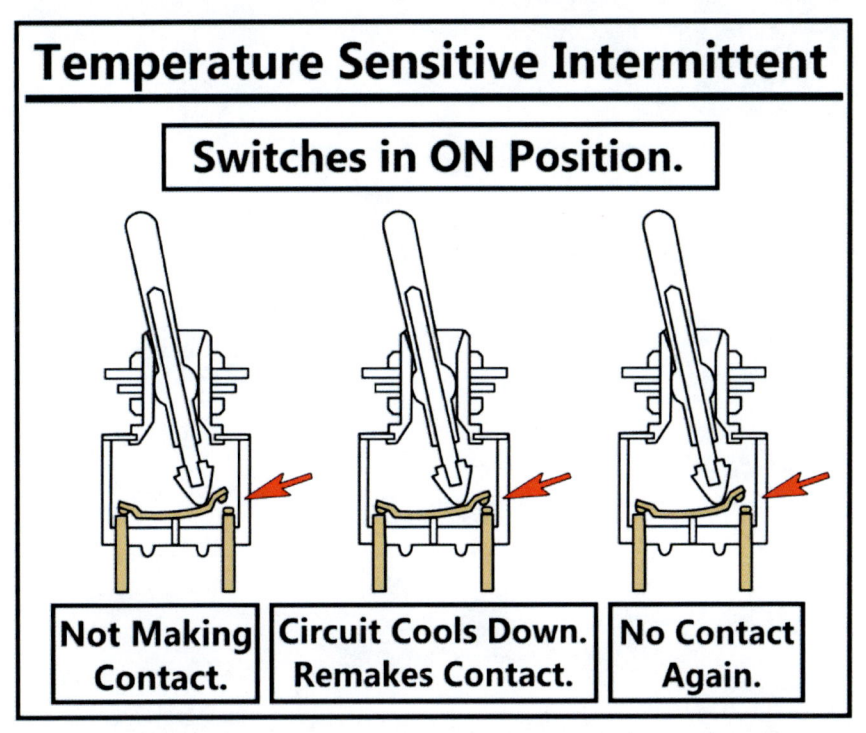

Temperature Sensitive Intermittent

Switches in ON Position.

| Not Making Contact. | Circuit Cools Down. Remakes Contact. | No Contact Again. |

(This is an example of what is happening inside the switch and how it is creating the intermittent electrical problem. The cause of the problem is a worn out switch contact. Many times, the internal contacts of a switch become worn and thinner than normal. When the circuit heats up, the moveable contact inside the switch expands. Because the contact is thinner than usual it expands more than it normally would because it heats up faster. This extra expansion causes the contact to break away from the stationary contact. When the circuit is allowed to cool back down, the moveable contact inside the switch shrinks back and re-establishes connection with the stationary contact. This switch is internally damaged and must be replaced to fix the problem.)

The Graphing Meter Approach: In another version of a temperature-related problem, the ON and OFF time of the circuit may be too fast to use open circuit testing. The intermittent may cause the circuit to almost have a flashing effect because of very short ON and OFF time. Normally with this symptom we might assume that the problem was vibration-related but after performing a quick wiggle test we might notice that there was no effect. If the problem cannot be reproduced with the wiggle test, then it might be temperature-related. One similarity between all fast acting intermittents is the use of the graphing meter. We must use our graphing meter once again just as we did before in order to effectively troubleshoot this problem.

Intermittent Example #2:

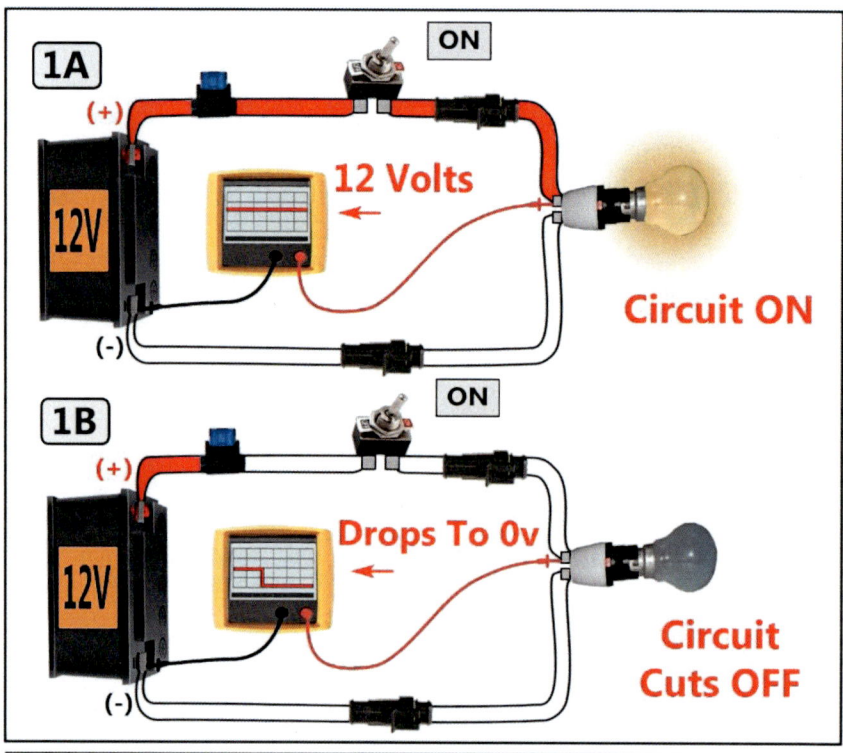

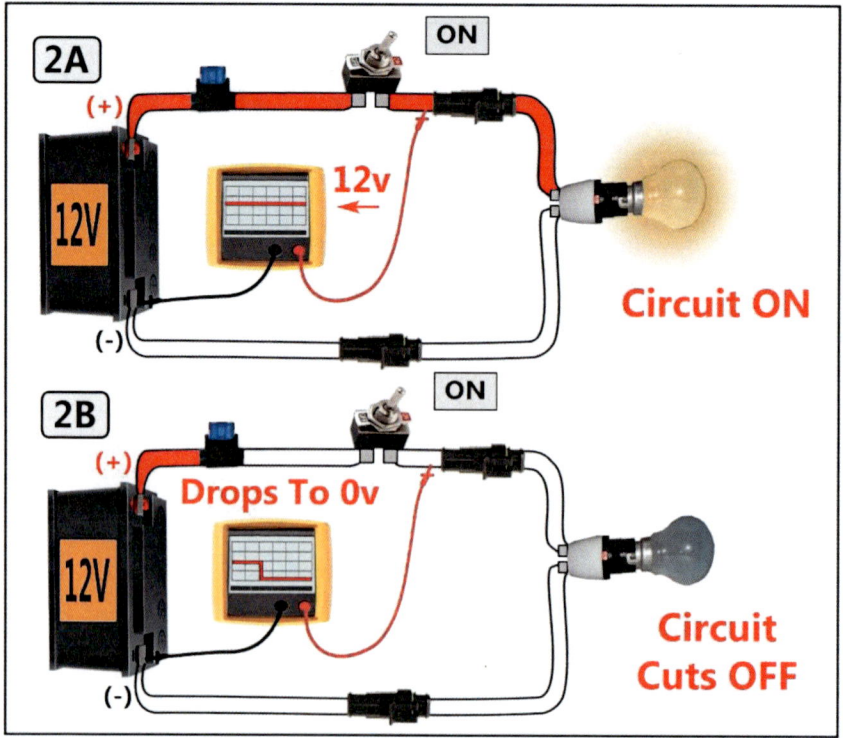

1A — 12 Volts — Circuit ON

1B — Drops To 0v — Circuit Cuts OFF

2A — 12v — Circuit ON

2B — Drops To 0v — Circuit Cuts OFF

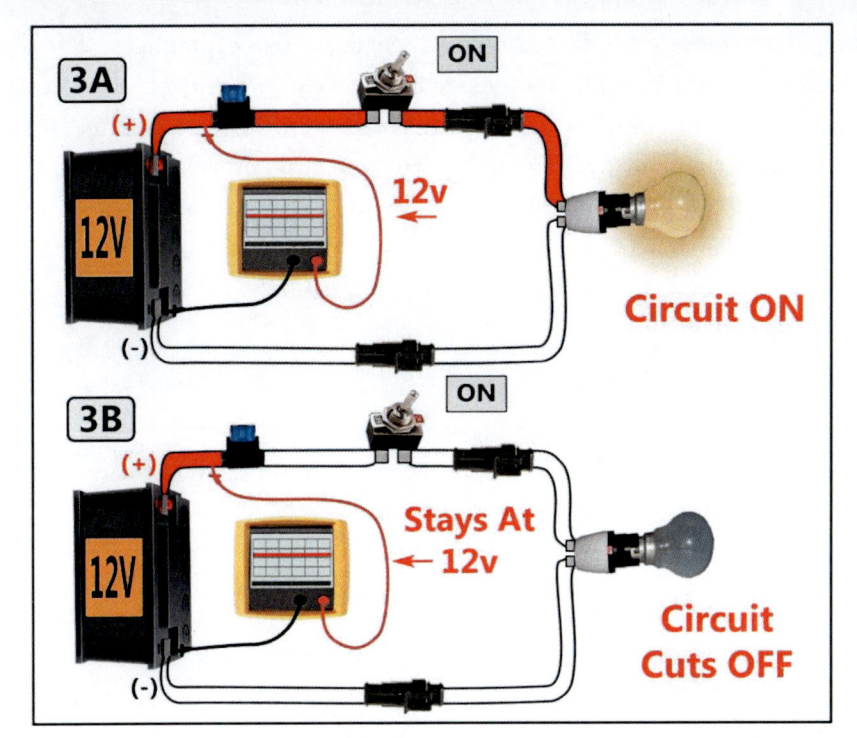

(In this version of our circuit we have a problem with a fast acting intermittent. The ON and OFF times only last about a few seconds. There are no vibrations around the circuit and Wiggle Test has no effect on the issue. We begin by using the graphing meter to test for voltage drop outs at various parts of the circuit to try to find the problem. The first two testing result illustrated indicate that the problem is not at the electrical device or at the connector since the voltage DID drop out during the intermittent cut off. The third illustration test results however remained at 12v during the intermittent cut off.)

This means that the problem must be somewhere between this test point and the last test point at the connector. The most nearby suspect part of the circuit we can begin to inspect is the fuse and the fuse holder. In this situation the cause of the problem was the fuse itself. After inspecting the fuse holder terminals, we pulled the fuse for visual inspection. We noticed that the fuse is actually broken at one end of the fuse filament yet still making contact enough to allow the circuit to turn ON.

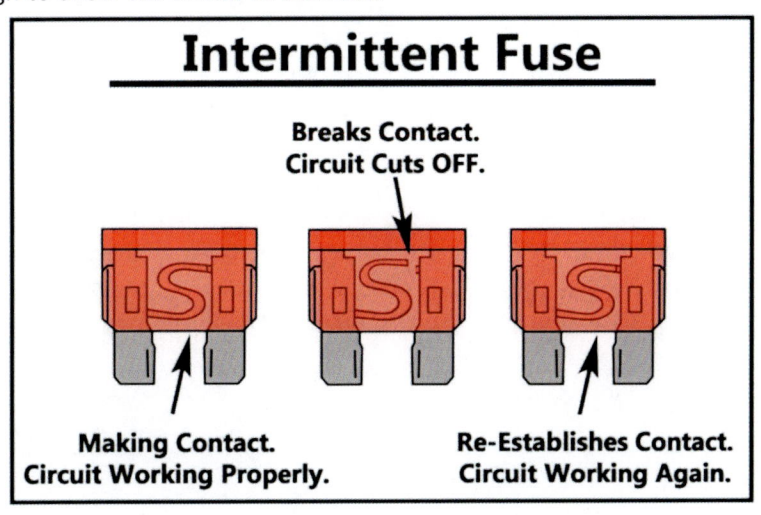

(In this example we see how the fuse could be the culprit to our temperature-related intermittent problem. The fuse filament inside the fuse was broken yet still making contact enough to allow the circuit to work. When the circuit was turned ON, the fuse filament would quickly heat up and expand away breaking the contact in the process. The circuit would then shut OFF. Then once the fuse cooled, the filament would re-establish contact immediately and the circuit would turn ON again. This would cause the light to turn ON and OFF almost like if it was flashing. The fuse was replaced and the problem was solved)

Never assume that a part like the fuse cannot be the cause of an intermittent electrical problem. This example may be a one in a million case but with intermittent problems always expect the unexpected. Inspect everything very closely as you come across them. You might just miss the problem while staring right at it. Intermittent problems are usually caused by something often overlooked in the circuit. Something you wouldn't expect to fail. Remain alert during testing and inspection.

Temperature Change Approach: In this method, we have to recreate the temperature change that causes the intermittent problem to happen in the first place. The method is easy to follow. We simply need to reproduce the hot or cold temperature changes that happen to the circuit. For reproducing the warm up stage, where the circuit cuts OFF, we can use a simple blow dryer. For reproducing the cool down stage, where the light turns back ON, we can either wait for it to cool or use non-flammable freeze spray to force cool the area.

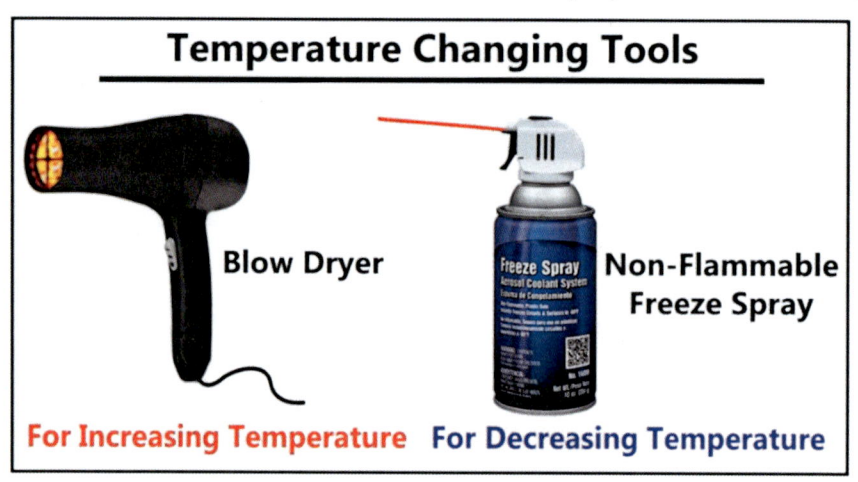

Temperature Changing Tools

Blow Dryer — For Increasing Temperature

Non-Flammable Freeze Spray — For Decreasing Temperature

(Tools used for forcing a temperature change in any electrical part of the circuit when testing for temperature-related intermittent problems.)

The goal is to use the temperature changing tools to attempt to force the circuit to cut ON or OFF. We will try to change the temperature of certain parts in the circuit to see if there is any effect to the circuit. We can do this to all the wiring and all parts of the circuit including **connectors, fuse terminals, relay terminals** (if equipped), **electrical device terminals, power source connections,** etc. Anywhere where there is an electrical connection on the circuit is a possibility for an intermittent contact issue.

Intermittent Example #3:

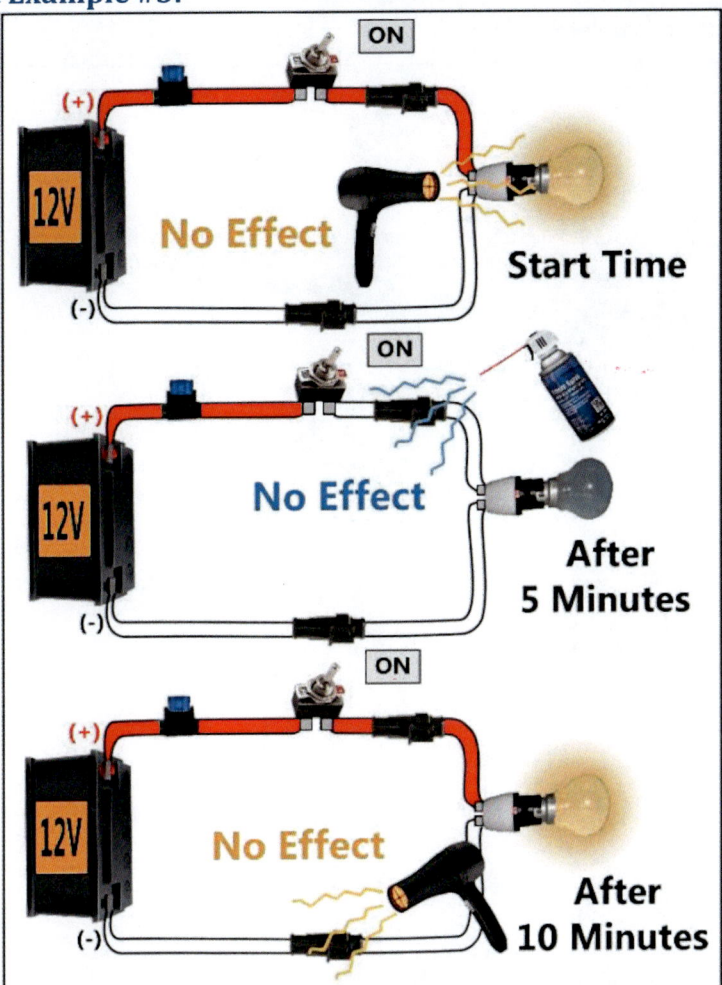

(In this illustration, we have a circuit that turns OFF after 5 minutes of working. Then after 10 minutes, the circuit cools down and turns back ON again. We have started by testing the intermittent circuit at different points using our temperature changing tools. We held the tools in position for about 30 seconds each to allow for a temperature change. We tested at the two connectors and the electrical device with no effect to the way the circuit normally cuts OFF. We must continue our spot testing somewhere else.)

Notice how when the circuit cuts OFF we switched our tool from blow dryer to the freeze spray. Then when the circuit cut back ON we switched back to the blow dryer. This was done so that we won't waste any time waiting for the circuit to turn ON by itself. We attempt to force the circuit either ON or OFF using the effects of temperature changing tools.

Fire Hazard Safety Notice: When using the blow dryer you want to move the tool around as you heat an electrical part. Dont focus the blow dryer on only one area for too long because you may begin to burn the area. Also, **NEVER** use the blow dryer around any FLAMMABLE objects or fluids as you can potentially cause a fire or explosion hazard. This includes the circuit's battery because the gases inside it are extremely flammable! It is okay to use on basic circuit parts but NEVER on any flammable objects. Please pay attention. You have been warned.

Let's continue now with temperature testing the other parts of the circuit to see if we can find the intermittent problem..

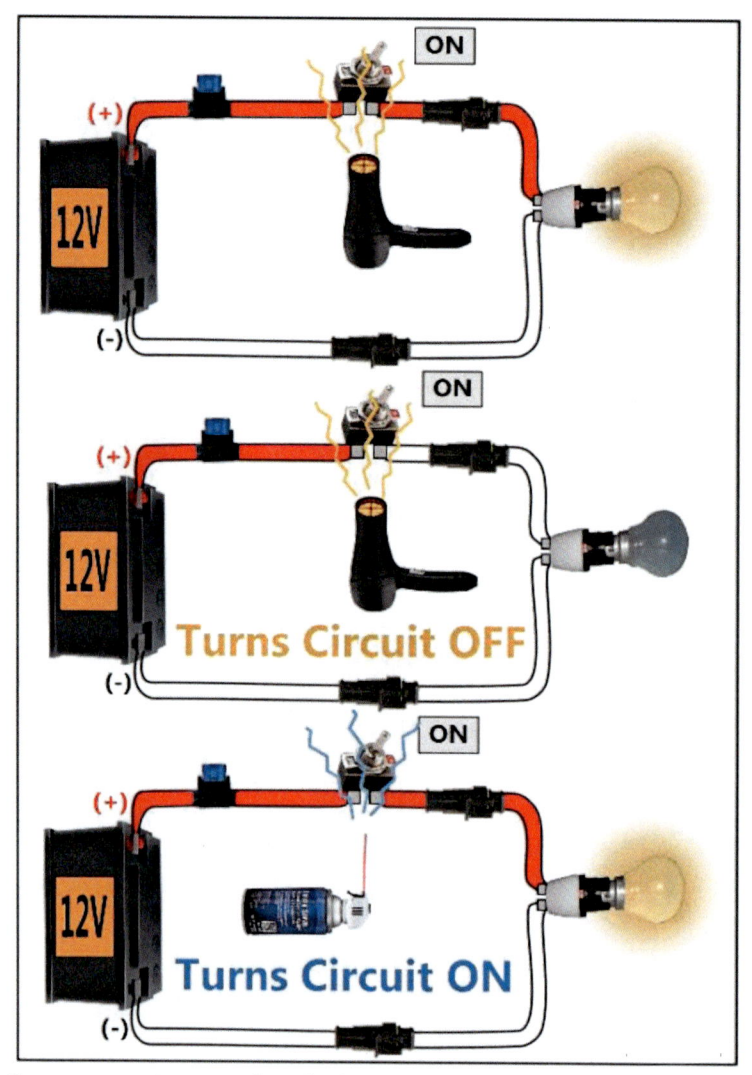

(In this illustration, we continue testing the intermittent circuit using our temperature changing tools. The next part to test is the switch. We held the tools in position for about 30 seconds each to allow for a temperature change. During the thirty seconds using the blow dryer, the light suddenly turned OFF. Cooling the switch using the freeze spray would also turn the circuit back ON. This temperature change in the switch changed the way the circuit normally cuts ON and OFF. Whenever the circuit cut OFF we could turn it back on using our freeze spray on the switch. This indicates an internal problem inside the switch. All that is left now is to inspect connections to the switch. If they are secure and tight, then the switch must be replaced to fix the intermittent problem.)

Remember that when dealing with any intermittent problem nothing is safe from being a possible cause. The cause of the problem could be anything. It could be a switch, the wiring, a connector, anything. Don't assume or discount any part of the circuit when dealing with intermittents. Use the test methods to best suit your problem and inspect carefully.

Ch.5: Relay Involved Intermittents:

In this chapter we will be looking at a more complex example of an intermittent problem that involves relay circuits. Remember than when dealing with relay circuits you always have at least two circuits involved. There is a coil side circuit and a switch side circuit. The relay is installed between these two circuits. If you are not familiar with how relays work or how to properly test them please refer to my other book "How to Test Relays and Involved Circuits" for a full explanation as well as helpful testing tips to use against them.

Reminder: Anything wrong with the coil side circuit will also affect the operation of the switch side circuit. In other words, the light bulb circuit can be prevented from working if there is a problem within the second circuit involved or if the relay itself is bad.

Since this problem involves two circuits, it can potentially take a very long time to diagnose unless we can test more efficiently. To accomplish this, we want to eliminate the possibility of it being a problem within the second circuit involved or in the relay. We will start our voltage testing at the relay. Let's see an example problem and how we would approach it..

Intermittent Example 1#:

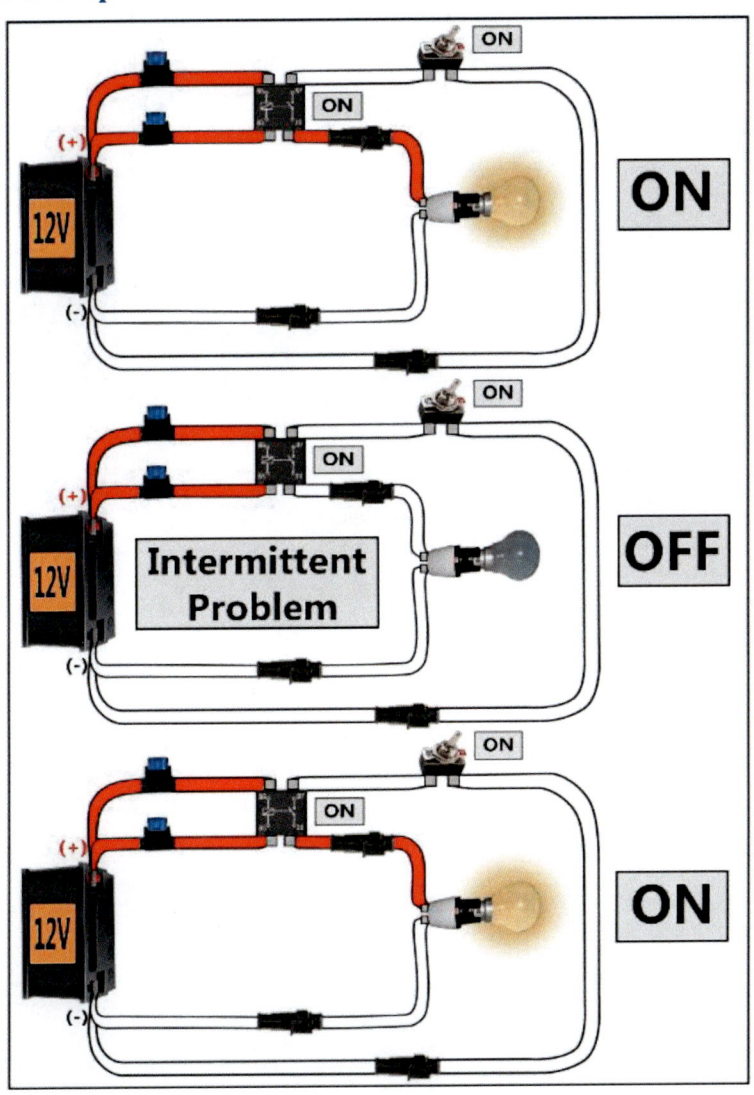

(In this example we now have a relay involved circuit with an intermittent electrical issue. The symptoms are that the problem starts whenever the circuit has been ON for a long enough time. Whenever the circuit is warmed up or after about 5 minutes, the light turns OFF temporarily. Then after about 10 minutes of waiting, the circuit cools back down and the light turns back ON suddenly. This problem has all the symptoms of a temperature-related intermittent.)

Since this problem usually takes some time before it begins to happen, we can use our Open Circuit Approach using our basic voltmeter to solve the issue. Let's see how it is done...

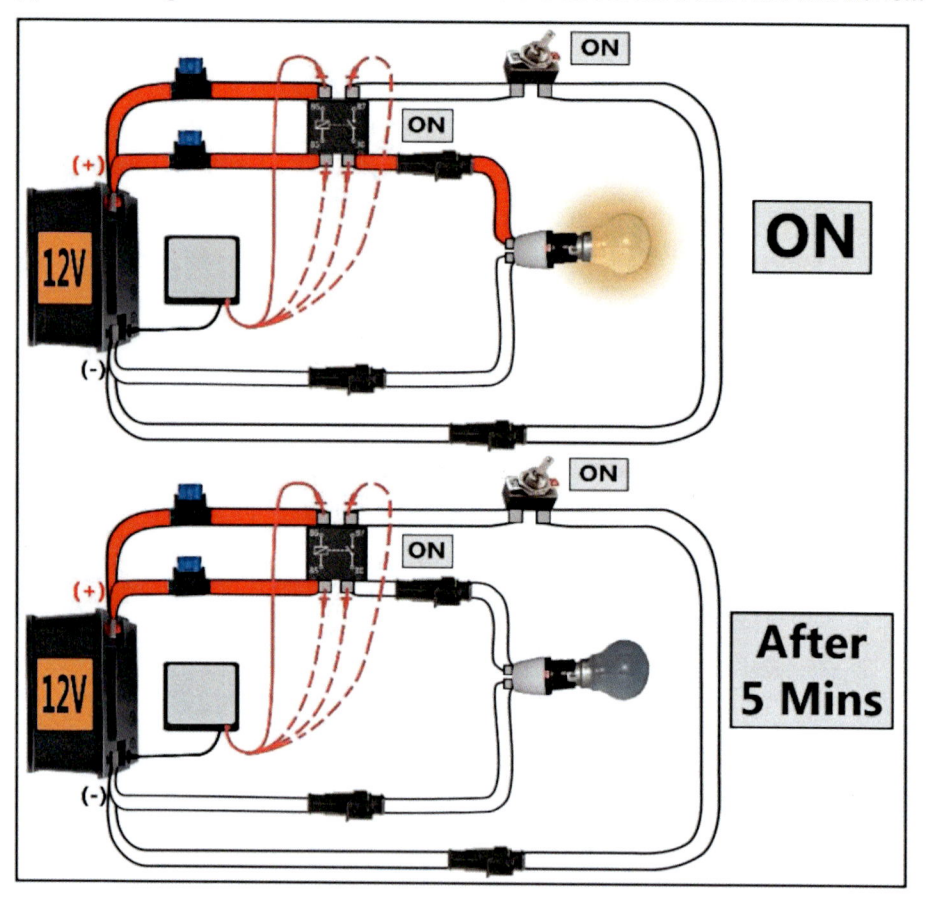

(In this approach we have started our voltage testing at the relay. We tested the voltages available to each of the relay terminals one by one with our red probe. We did this during the time the circuit was ON as well as when the circuit was OFF.)

We recorded each voltage value at the terminals to analyze and note if there is a problem with the relay. Here are the results that we found...

Relay Terminal Guide:

Test Results

Normal Operation. Circuit ON Time.

During Intermittent. Circuit OFF Time.

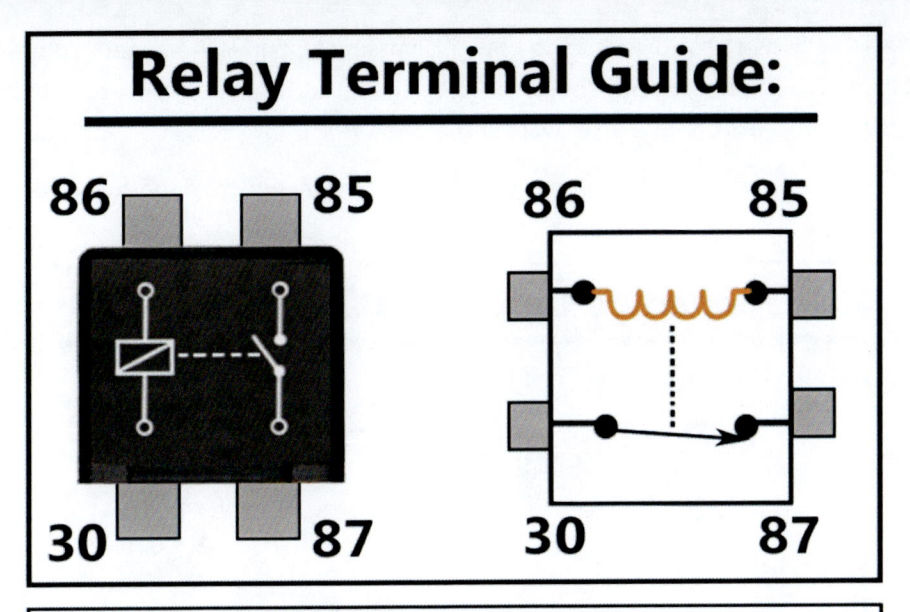

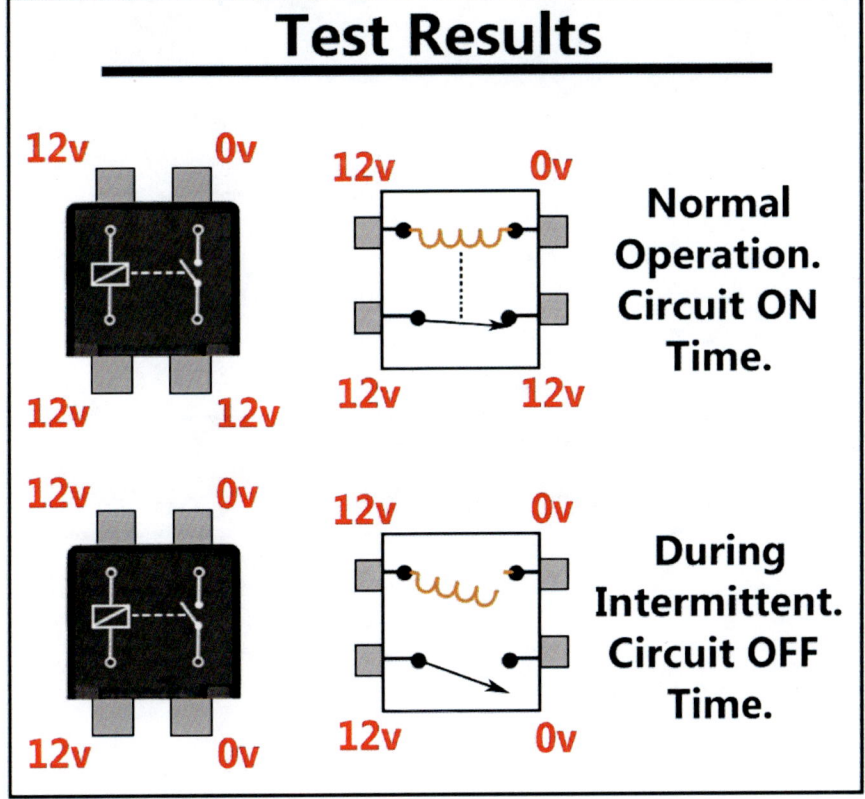

(In this illustration, we show the test results we collected during relay testing. During the circuit's normal operation time, the voltages were normal. We had 12v at pin 86 (good), 0v at pin 85 (good), 12v at pin 30 (good) and 12v at pin 87 (good). When the intermittent problem turned the circuit OFF we had 12v at pin 86 (good), 0v at pin 85 (good), 12v at pin 30 (good) and 0v at pin 87 (bad). The 0v reading on pin 87 means that there is a problem at the relay.)

The change in voltage at pin 87 after the circuit cut OFF means that something is happening at the relay when the intermittent problem started. The relay switch is not letting the voltage pass through it to the light bulb when the intermittent problem is in effect.

There are actually two possibilities as to how this missing voltage can happen. You can either have a bad switch inside the relay that is losing contact after warm up or a problem in the coil of the relay that breaks contact after warm up.

We will add a clamp on ammeter to the second circuit to see which of the two possibilities it is.

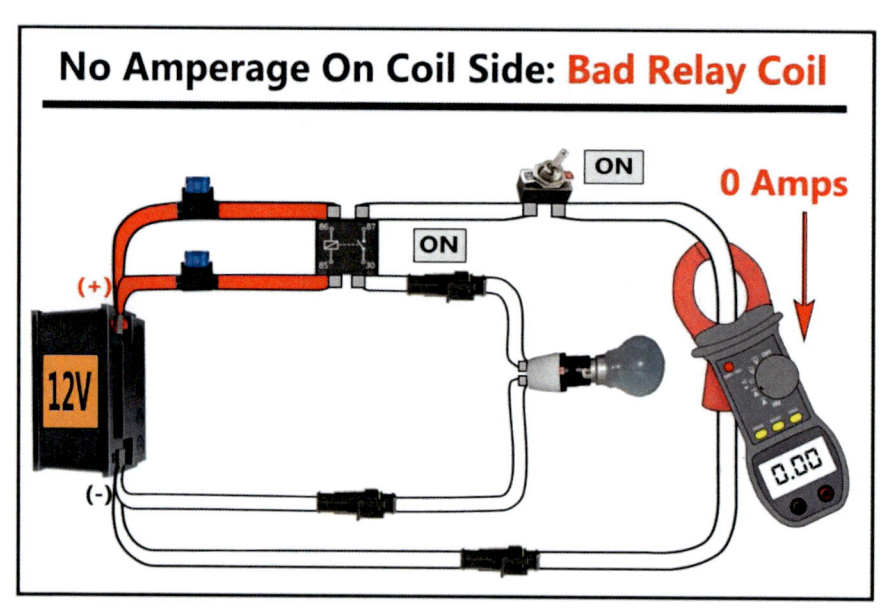

No Amperage On Coil Side: Bad Relay Coil

(In this illustration we have added a clamp-on ammeter to the second circuit (the coil side circuit) to check if it is working during the intermittent. The reading is 0 amps, meaning the coil circuit is also NOT working during the intermittent. This confirms that the cause of the intermittent problem is the coil inside the relay.)

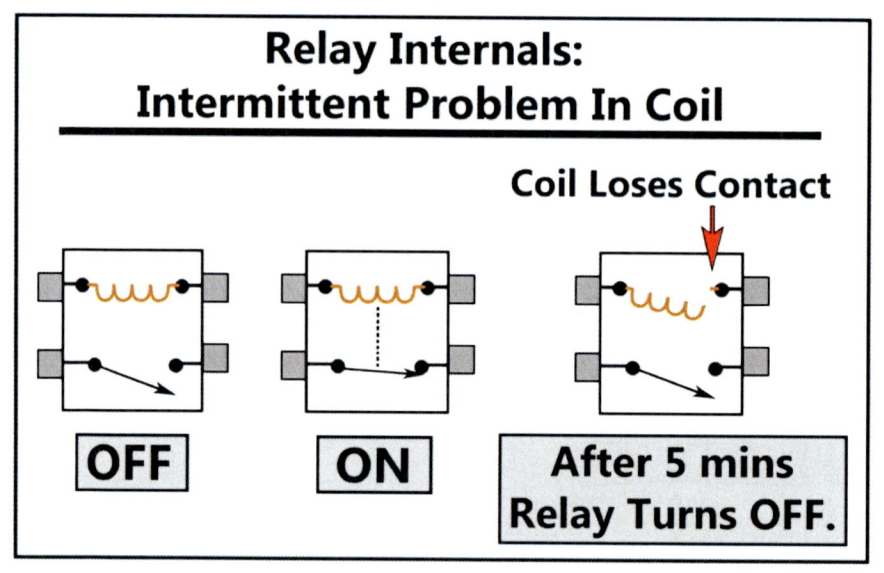

**Relay Internals:
Intermittent Problem In Coil**

Coil Loses Contact

OFF | ON | After 5 mins Relay Turns OFF.

(This is image shows the source of the intermittent problem in our last example. The relay coil was internally broken yet still making contact enough to allow the circuit to work.)

When the circuit would turn ON, the broken coil would warm up and expand away, breaking contact in the process. This would cause the light circuit to cut OFF as well. The relay in this case is internally damaged and needs to be replaced in order for the problem to be fixed.

This intermittent relay is surprisingly more of a common problem than one may think. Usually the cause of the internal damage can be due to age, extended long periods of ON time and even manufacturing defects. Don't assume that a relay can't be cause to an intermittent problem. Everything in the circuit is a suspect.

Remember that any problems on the coil side adversely affects the switch circuit as well. If the problem cannot be found after testing at the relay, the problem can either be in the coil side circuit or the switch side circuit. As long as you remember previous voltage training, two circuits shouldn't be a problem. Just take on each circuit one at a time. Next, analyze the readings you take and understand where in the circuit you need to move your probes to next.

Let's look further into another complex example involving relay circuits..

Intermittent Example 2#:

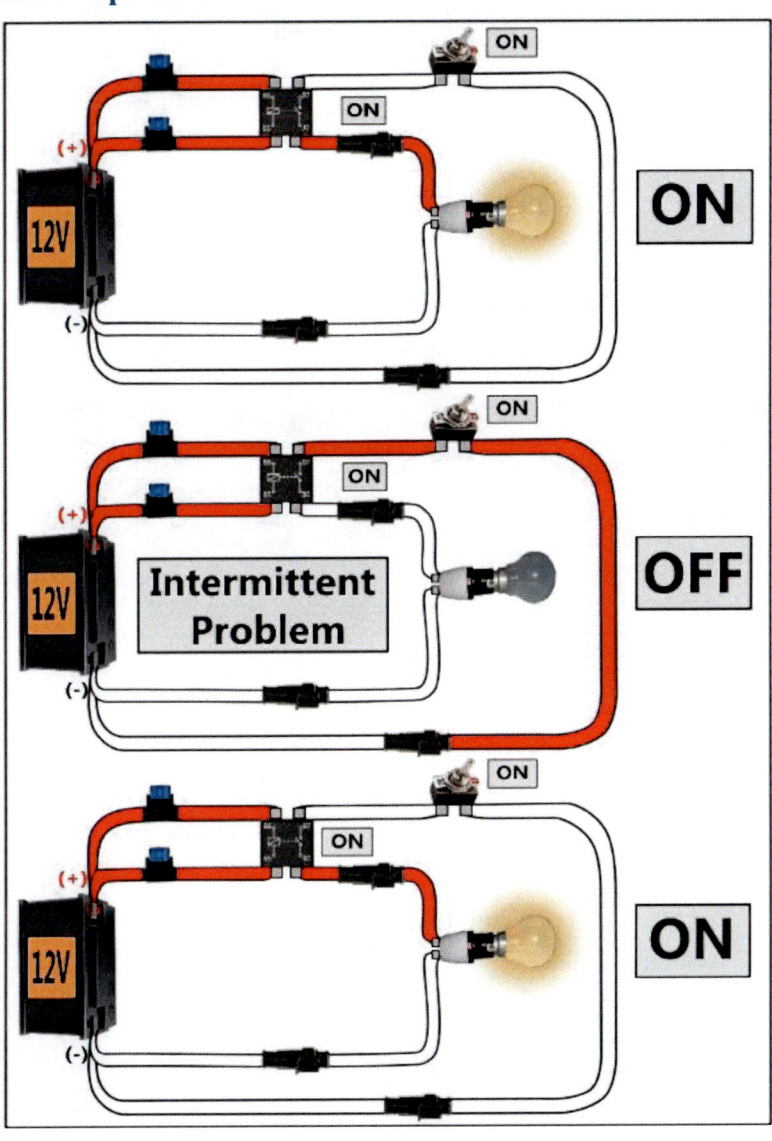

(In this example we have two relay involved circuits with a different intermittent electrical issue. The symptoms are that the light turns ON and OFF randomly. Neither temperature nor time make a noticeable difference to the random cutting ON and OFF of the light circuit. This problem has all the symptoms of a vibration-related intermittent.)

The cutting ON and OFF of the circuit are random and seemingly unpredictable. This is a common symptom of a vibration-related intermittent. In order for us to track this problem down we will have use the graphing meter test while wiggling the wiring and all components of the circuit. The goal is to be able to reproduce the problem while voltage testing. As done before, we will begin our testing at the relay. Let's take a look at the steps to finding this problem...

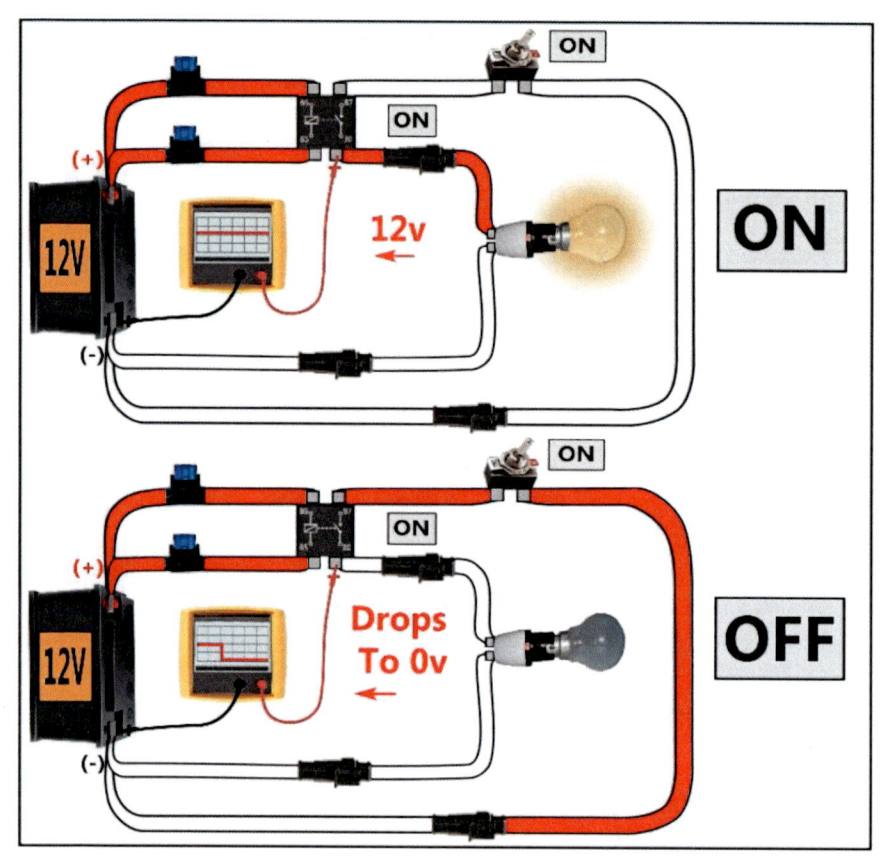

(In this illustration we start our voltage testing at the relay terminals. We begin by monitoring the voltage at relay terminal #87 (switch side circuit). We shake all the wiring and parts of the circuit attempting to reproduce the vibrations that cause the problem. With the relay ON the meter should normally read 12v at terminal #87 if everything is working properly. During the ON part of the intermittent, the graphing meter reads 12v (good). When the intermittent cuts the circuit OFF, the meter reading fell to 0v (bad). This means the voltage is being lost across the relay or before the relay. We are on the right track but must continue voltage testing to get closer to the problem.)

The voltage drop out indicates that the problem is either in the relay, on the other power wire that goes to the relay or in the coil side circuit. Let's test at the other switch side terminal of the relay to eliminate each possibility.

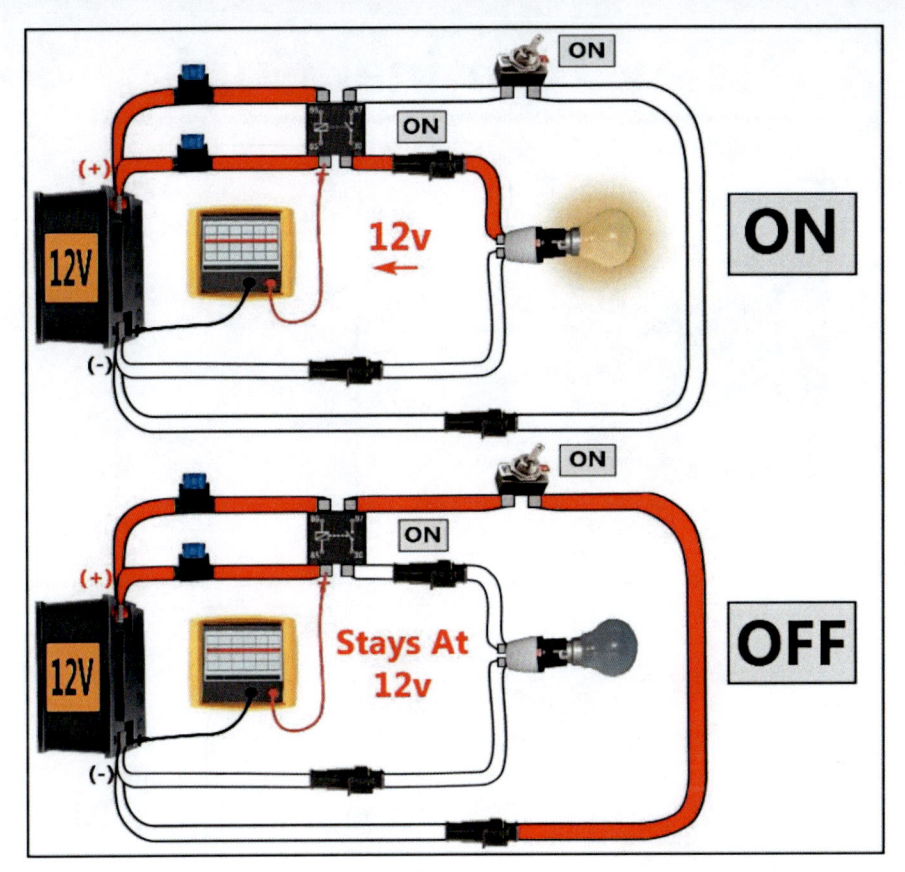

(In this illustration, we continue voltage testing at the other switch side relay terminal. We are now monitoring the voltage at relay terminal #30 (switch side circuit). We shake all the wiring and parts of the circuit attempting to reproduce the vibrations that cause the problem. With the relay ON, the meter should normally read 12v at terminal #30 if everything is working properly. During the ON part of the intermittent, the graphing meter reads 12v (good). When the intermittent cuts the circuit OFF, the meter still reads 12v (good). This reading indicates that nothing is wrong with the power wiring on the switch side circuit going back to the battery.)

The problem now is either the internal switch inside the relay or the something in the coil side circuit. We will have to check to see which of the two possibilities it is by testing for voltage at coil side terminals of the relay.

Relay Terminal Guide:

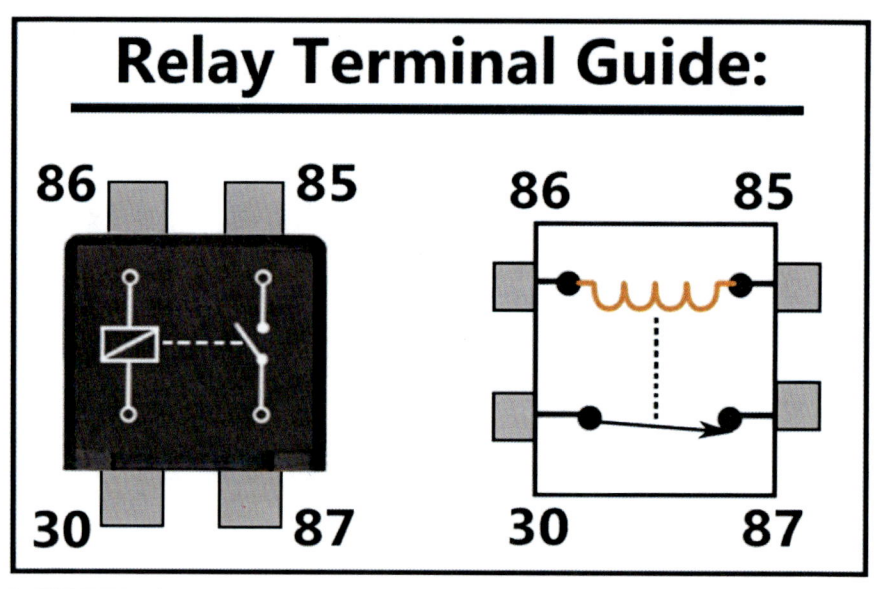

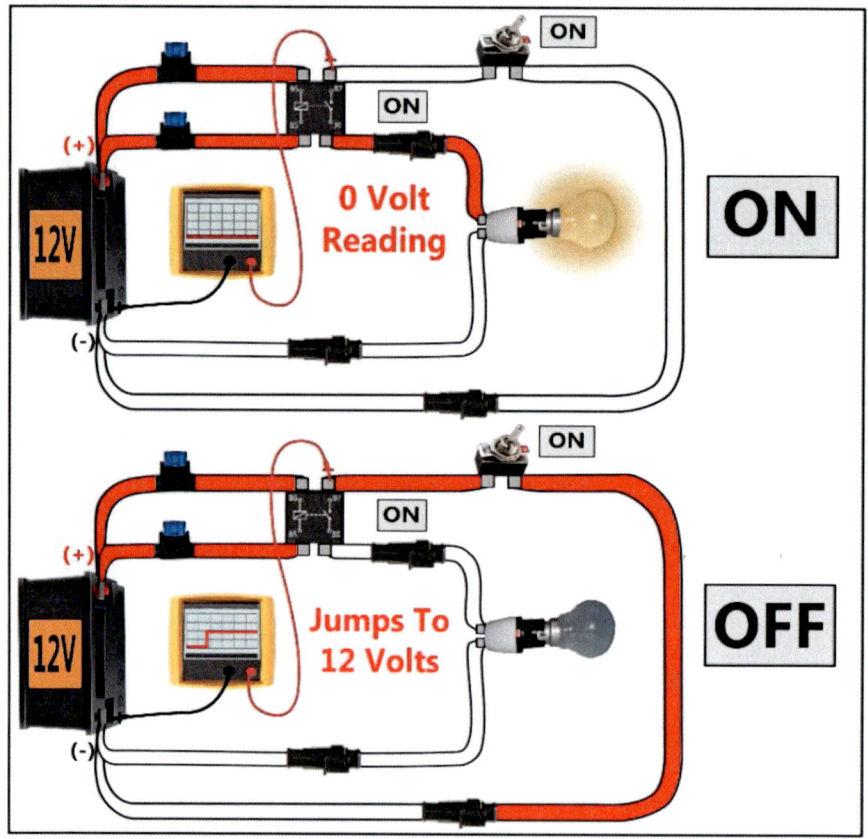

(In this illustration we continued voltage testing at the coil side relay terminal to eliminate one of the two possible causes. We are now monitoring the voltage at relay terminal #85 (coil side circuit). We shake all the wiring and parts of the circuit attempting to reproduce the vibrations that cause the problem. With the relay ON, the meter should normally read 0v at terminal #85 if everything is working properly. During the ON part of the intermittent, the graphing meter reads 0v (good). When the intermittent cuts the circuit OFF, the meter immediately jumped up to 12v (bad). This reading indicates that something is happening on the coil side circuit that is prevent the light from turning on.)

Note: This reading also means that the internal coil or switch inside the relay is NOT part the problem. The reason why is due to the fact that voltage was able to make it from one side of the relay coil to the other side.

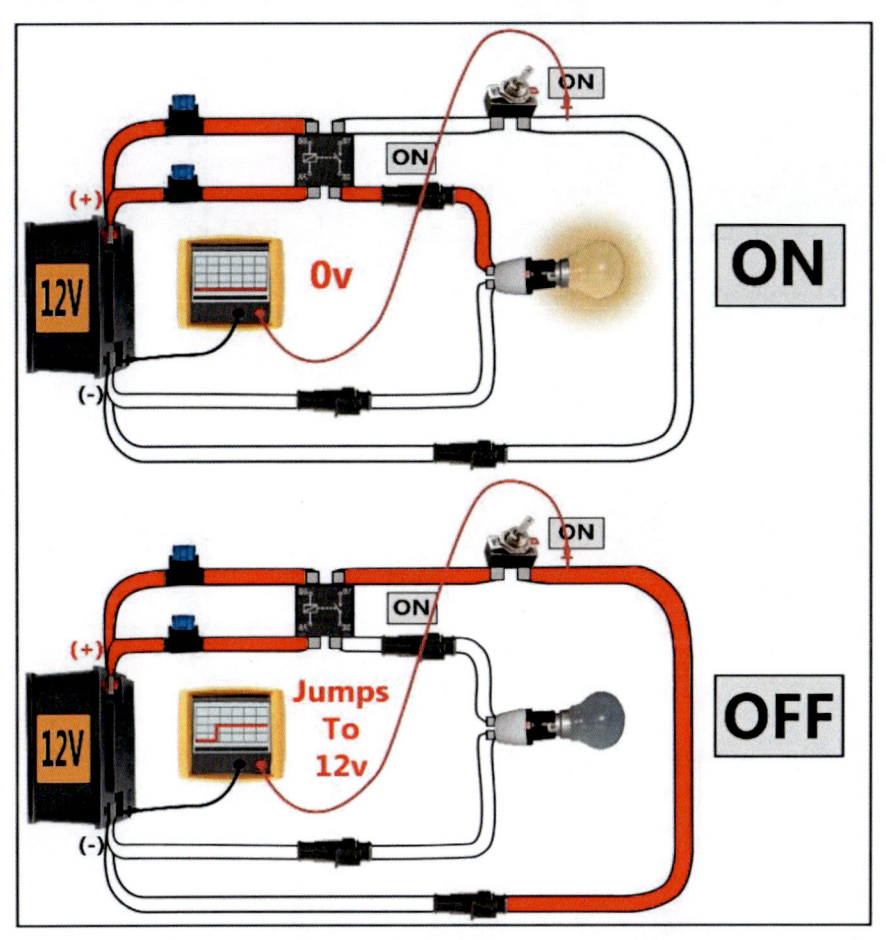

(In this illustration, we continued testing at the electrical switch on the coil side circuit. At this test point, we are now monitoring the voltage that is present after the switch. We shake all the wiring and parts of the circuit attempting to reproduce the vibrations that cause the problem. With the relay ON, the multimeter should normally read 0 volts at this point if everything is working properly. During the ON part of the intermittent, the graphing meter did in fact read 0 volts (good). When the intermittent cuts the circuit OFF, the meter still jumped up to 12 volts (bad).These reading indicate that the problem is still more down the negative line on the coil side circuit.)

With these reading at this test point, we can also assume that the electrical switch is working properly and is NOT cause of the intermittent problem.

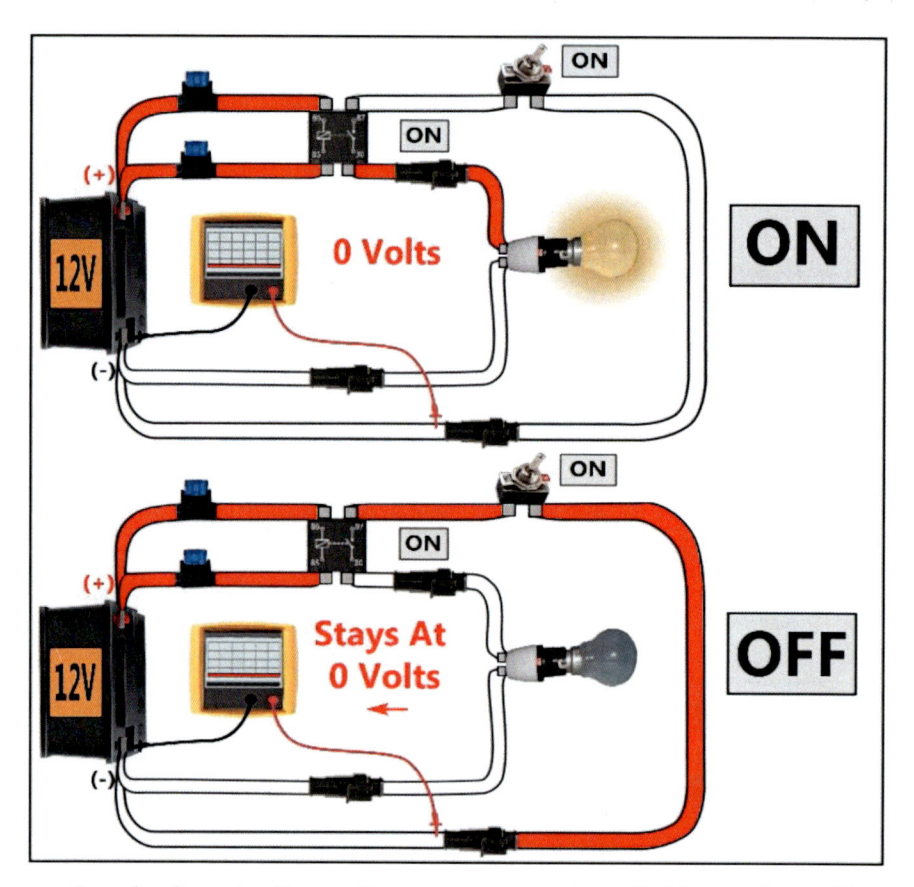

(Here we continued voltage testing at the connector on the coil side circuit. At this test point, we are monitoring the voltage that is present after the connector. We shake all the wiring and parts of the circuit attempting to reproduce the vibrations that cause the problem. With the relay ON, the meter normally reads 0v at this point if everything is working. During the ON part of the intermittent, the graphing meter reads 0v (good). When the intermittent cut the circuit OFF, the meter reading stayed 0v (good). This measurement indicates that the problem must be between this test point and the previous test point.)

The connector is the first suspect in the area so we will inspect it thoroughly. After careful inspection of the connector we notice that whenever we move this connector, the circuit turns ON or OFF. The connector on the coil side circuit was the source of this intermittent problem that would turn the light ON or OFF randomly. In order to fix this problem, the connector must be replaced.

Intermittent Problem: Loose Connector Terminal.

Loose Terminal Inside Connector

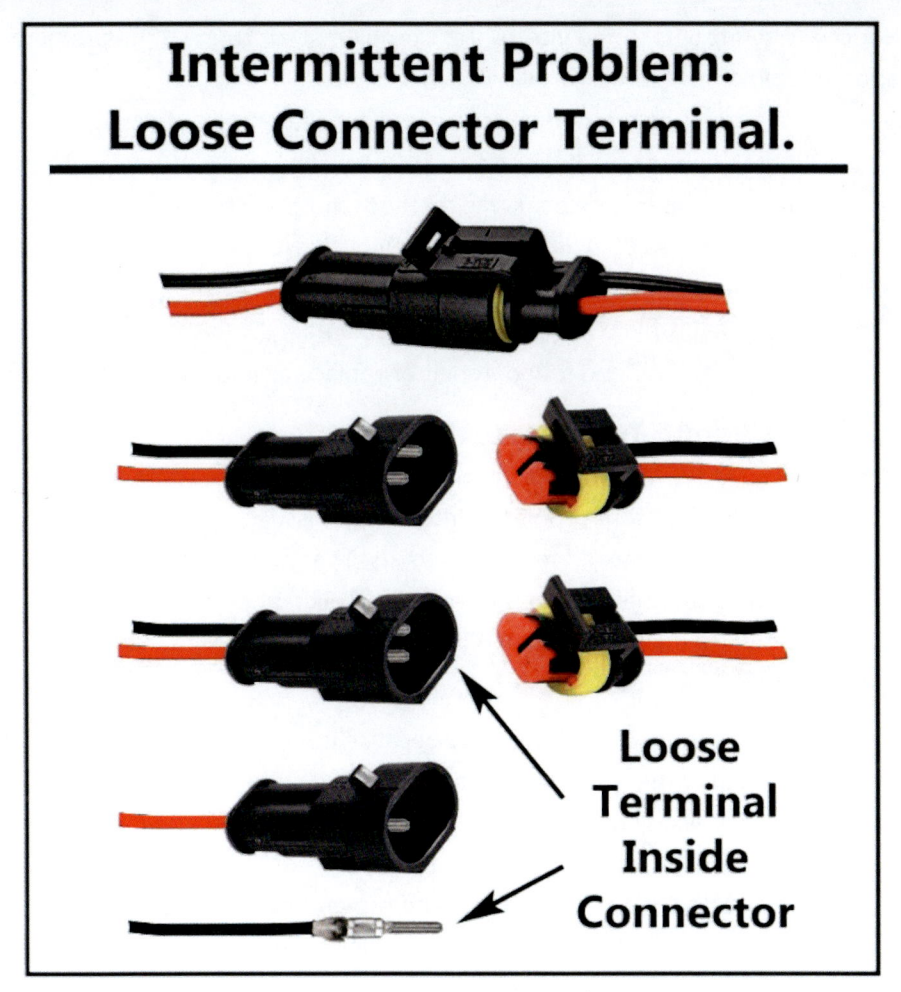

(This image shows the inside of a two wire connector that was causing an intermittent electrical problem. One of the terminals inside the male connector was loose inside the case of the connector and caused intermittent contact with the female terminal once it was installed. The repair required that the entire connector be replaced in order to secure the problem would not return.)

As you can see, the true cause of the intermittent problem in our relay involved circuit was a simple loose terminal inside the connector's plastic casing. Often these plastic cases are damaged or wear out over time. At that point they have the potential to become a problem down the road.

This damaged connector caused the coil inside the relay to turn OFF intermittently and as a result would also turn OFF the switch side circuit's light bulb. Always remember that the two circuit are connected to each through the relay. If something happens to the coil side circuit it WILL affect the switch side circuit as well.

By now you should see a trend in the testing methods that we would use to solve an intermittent problem. The methods for finding the source of the problem are generally the same regardless of the type of circuit you have. Let's recap the basic steps on diagnosing any intermittent problem.

- #1: Notice the symptoms to the intermittent problem. Determine if the problem is vibration-related or temperature-related.
- #2: Select the best testing method to use depending on the type of situation.
- #3: Take note of the meter readings and assess what they mean. When shaking the wiring or using the temperature changing methods, notice the changes you may create at each test point you observe. Always inspect previously worked on or newly soldered areas first.
- #4: Narrow in on the problem using your collected data.
- #5: Once the problem has been found, repair or replace as necessary to fix the problem.

Important Fact About Any Diagnostic Technique:

The test methods I have provided are designed to allow you to have some sort of system or plan of attack against intermittent problems. Intermittent issues are by no means easy to solve as they require more advanced testing methods. These methods WILL work to find the problem but sometimes you may notice that you didn't even need to use any testing after simply giving everything a quick inspection. That is okay too and it WILL happen many times while diagnosing.

A good pre-testing inspection is very important because it might save you a lot of time and trouble when trying to find the source of the problem. Always inspect the circuit first at potential problem areas. Many times you may be able to find that a lot of electrical problems could have been solved faster by simply wiggling and inspecting the connections and terminals of the circuit. Do the easier stuff first and remember to perform thorough inspecting.

Ch.6: Miscellaneous Tips and Tricks

Thank you very much for your purchase of this book. I honestly strive to make each of my books a valuable experience for all my readers. On the next installment of the "How To Test Like A Pro" series, I will show you how you too can solve problems involving Electrical Sensors and Control Module Circuits. I have added this last chapter to my book to include some extra tips and tricks for you to take along with you on your electrical journey. Thank you again and Good Luck!!

Important Note About Having a Good Meter Probe Connection:

During testing, if the meter probes are not connected securely, you might end up with bad meter readings. You may think there is a problem at a test point in the circuit after seeing 0v, when in fact the meter probes are not connected properly. A loose or poorly connected meter probe will cause a voltage drop out or 0v reading on the meter. These are misguided readings and can lead to a lot of confusion when diagnosing. Always make sure you have a secure meter probe connection before taking note of any readings.

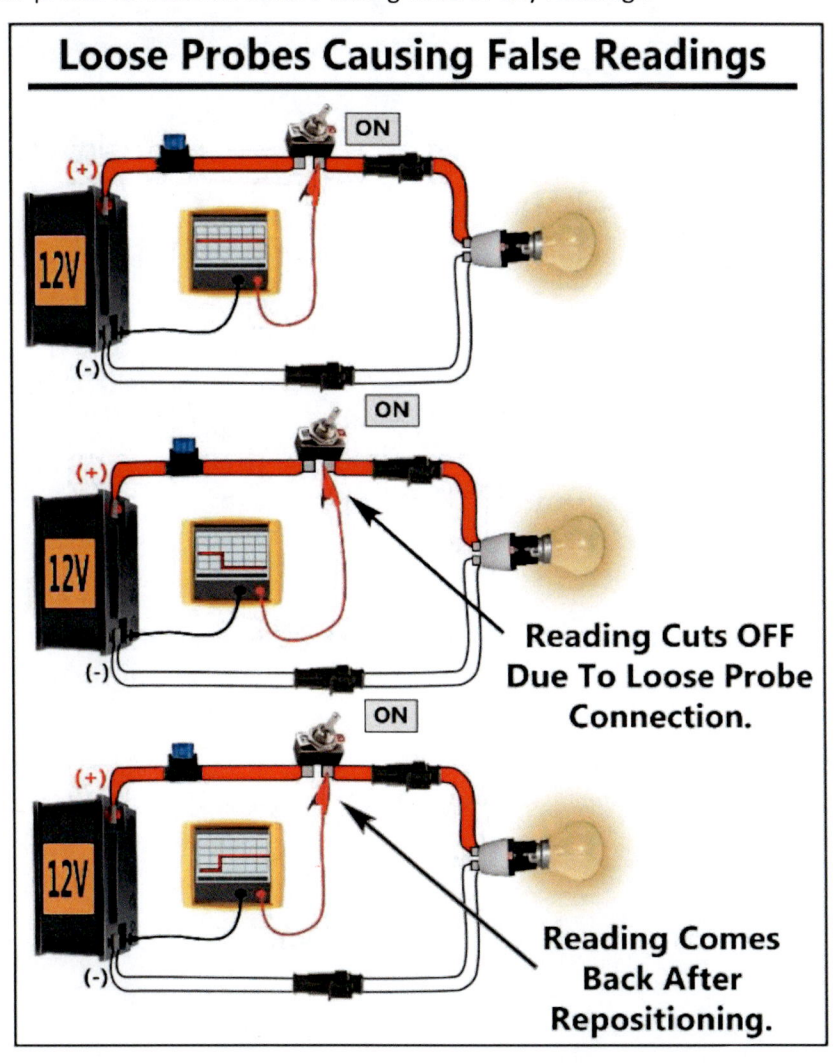

Loose Probes Causing False Readings

Reading Cuts OFF Due To Loose Probe Connection.

Reading Comes Back After Repositioning.

(Example of how bad test probe connections can cause a false reading if not installed securely. We have used the alligator type meter probes to demonstrate this issue. To prevent this, always make sure your test probes are secure and tight on the test point that you are diagnosing at.)

Using Metal Frames As A Ground:

In modern electrical circuits, sometimes instead of using negative wire, the engineers will choose to save wiring by using structural frame pieces near the circuit. They use this conductive metal frame piece nearby to complete the negative side of the circuit without the need of extra wiring. The basic idea is to run the power carrying wiring to the electrical load and then use metal frame structures as a negative path. The engineers usually accomplish this by attaching a short wire to one side of the metal frame and then using another short wire on the other end of the metal frame to connect back to the negative battery terminal. The metal frame may extend a large distance from the battery to the electrical load. This design method allows a circuit to be able to extend longer distances away from the power source without having to worry about extra ground wiring costs.

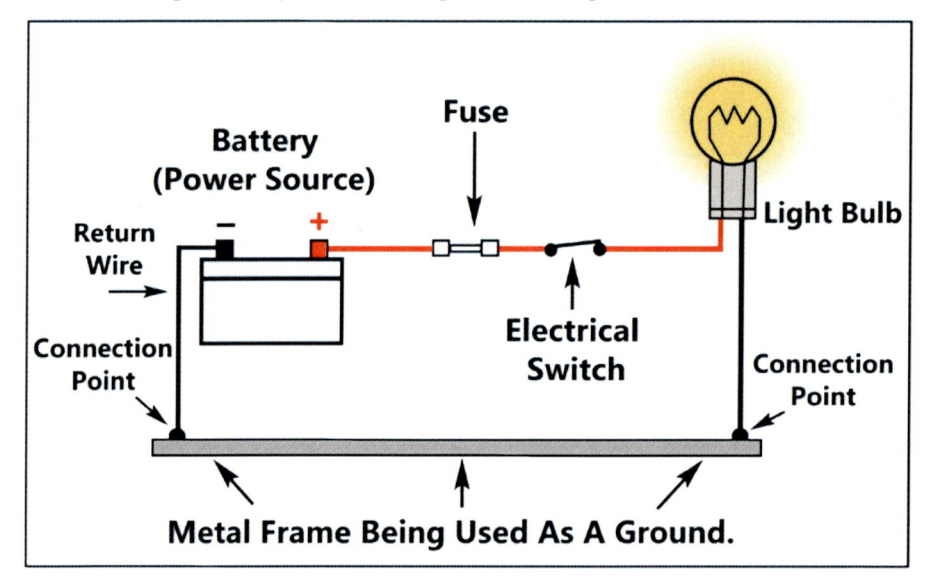

(Illustration of a circuit that uses a metal structure frame to the complete the circuit on the negative side.)

Now this design change generally shouldn't present a problem because the structural frame pieces used are often very sturdy and not damaged easily. But if for some reason there was frame or structural damage, the circuit CAN potentially be broken and not work anymore. In this case you would perform open circuit testing just as you have before in other examples only now you factor in the frame being used as a negative side.

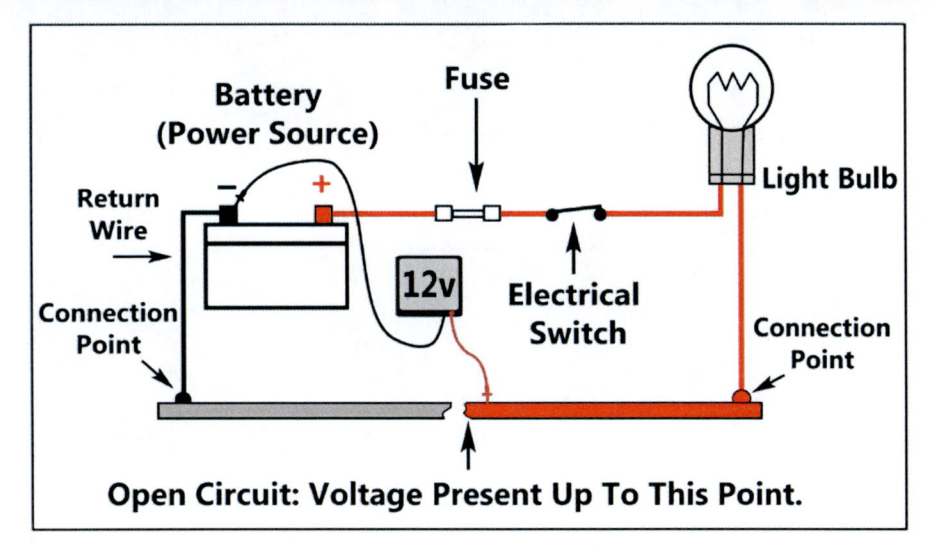

Battery (Power Source) — **Fuse** — **Light Bulb** — **Electrical Switch** — **Connection Point** — **Return Wire** — **Connection Point** — 12v

Open Circuit: Voltage Present Up To This Point.

(Example of structural frame damage being the source of an open circuit. Testing methods remain the same for open circuit troubleshooting.)

Loose Screw-Type Ground Connections:

Another common fault on circuits that use structural frames as a negative side is when the connection points to the frame become worn or loose. The common types of connections used are often the screw type connections. These connections can become worn or loose over time. Continuous vibrations near these connection points are often enough to steadily begin to back the screw off a thread or two. Over time they can become so worn or loose that they begin to create resistance problems and even intermittent electrical issues.

Loose Screw Type Connections. Cause Of Intermittent Problems.

(This is a visual representation of how the typical screw-type ground connections are installed on a frame. Over time they may become loosened through nearby movement or vibrations and begin to lose contact. This can eventually develop into an intermittent issue. Make sure to check the voltage across these points to make sure they are not the problem)

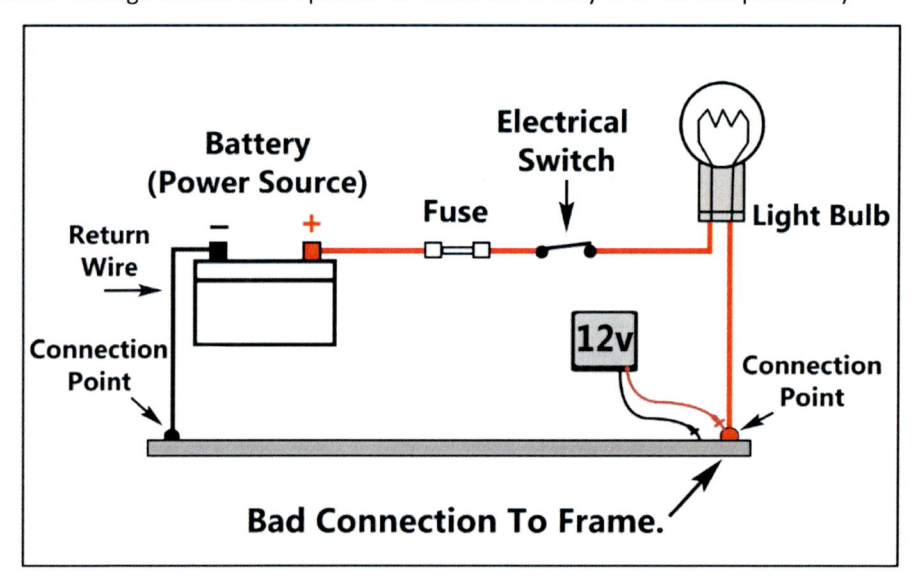

(In this illustration we are testing the connection points for the frame. The goal is to place one test probe on the screw connector and the other probe to the frame. This will tell you the voltage difference between the two points. Ideally there should be no voltage drop across a connection. In this example our meter read 12v indicating a bad connection point. Repair or replace as necessary)

Cooling or Ventilation Systems Causing Electrical Problems:

Usually when an engineer puts a lot of circuits together in one place there is some sort of ventilation system or air conditioning system to keep the circuits cool. As you have seen, temperature can play a major part in an electrical problem. Many times electrical parts will be rated to work best up to a certain maximum temperature. This is known as the max operating temperature. If the temperature of these electrical parts go over this limit, they will begin to work improperly or not at all.

If the circuits being tested are in a closed environment, it is necessary to have some sort of ventilation or cooling to prevent electrical parts from overheating. If the cooling system fails it will cause the circuits to begin to work improperly or shut OFF completely, due to heat. Before jumping to conclusions over symptoms always check to make sure that if there is a cooling or ventilation system for the circuit, that it is ON and working as designed.

Conclusion:
I congratulate you on making it this far in the Part 2 of this series. In Part 3 of the series I will pick up where I left off and show you more on how you can solve very complex electrical problems using your meter. I will be focus more on electrical sensors and control module problems in the next installment. Stay tuned on "How To Test Like a Pro".

Made in United States
North Haven, CT
24 April 2023